Feeding and Survival Strategies of Estuarine Organisms

MARINE SCIENCE

Coordinating Editor: Ronald J. Gibbs, *University of Delaware*

Recent Volumes:

A Continuation Order Plan is available for this series. A continuation order will bring delivery of each new volume immediately upon publication. Volumes are billed only upon actual shipment. For further information please contact the publisher.

Library

CHESTER COLLEGE

This book is to be returned on or before the last date stamped below, unless it is recalled for the use of another borrower.

Feeding and Survival Strategies of Estuarine Organisms

Edited by
N.V. Jones
University of Hull
Hull, United Kingdom

and

W. J. Wolff
Rijksinstituut voor Natuurbeheer
Den Burg-Texel, The Netherlands

PLENUM PRESS • NEW YORK AND LONDON

Library of Congress Cataloging in Publication Data

Main entry under title:

Feeding and survival strategies of estuarine organisms.

    Proceedings of a joint meeting of the Estuarine and Brackish – Water Science Association, the Nederlandse Oceanografenclub, and the Hydrobiologische Vereniging, held at the University of Hull, Hull, U.K., Sept 17 – 20, 1980.
    Includes bibliographies and index.
    1. Estuarine ecology—Congresses. 2. Estuarine fauna—Food—Congresses. 3. Estuarine flora – Congresses. 4. Plants—Nutrition—Congresses. I. Jones, N. V. II. Wolff, W. J. (Wim J.) III. Estuarine and Brackish – Water Sciences Association. IV. Nederlandse Oceanografenclub. V. Hydrobiologische Vereniging.
QH541.5.E8F43   574.5'26365   81-12005
ISBN 0-306-40813-9   AACR2

Proceedings of a joint meeting of the Estuarine and Brackish-
Water Sciences Association, The Nederlandse Oceanografenclub
and the Hydrobiologische Vereniging, held at the University
of Hull, Hull, U.K., on September 17 – 20, 1980

© 1981 Plenum Press, New York
A Division of Plenum Publishing Corporation
233 Spring Street, New York, N.Y. 10013

PROLOGUE

In September 1978 the Estuarine & Brackish-Water Sciences
Association and the two Dutch organisations, the Netherlands Oceanografenclub and Hydrobiologische Vereniging, arranged a joint meeting
entitled "The Environmental Impact of Large Hydraulic Engineering
Projects". That meeting was held at Middelburg in the Netherlands
and the papers published in the Hydrobiological Bulletin, Vol. 12,
Nos. 3/4 in December 1978.

This present volume is the result of the return meeting which
was held at the University of Hull in September 1980. The broad
title was deliberately chosen to allow discussion of strategies
recognised, or looked for, in various groups of organisms. This
approach is reflected in the subjects of the papers and in the inter-
pretations adopted by different authors.

The main aim of the meeting was to try to understand how estuarine organisms of different sorts manage to make a living. It is
clear that they need adaptations in order to survive in estuaries
but they may need strategies in order to make a living. It was
easily forseen that the interpretation of the title and the recog-
nition of strategies would vary considerably depending on which
organisms the various authors are concerned with. The authors were,
however, allowed freedom to select the approach they considered most
appropriate for their organisms and their data. This has resulted
in 21 diverse papers on organisms varying from bacteria to birds and
that vary in their approach from descriptive biology to hypothetical
discussions and useful reviews.   There is, clearly, considerable
variation in the state of our knowledge of the different taxonomic
groups which is, no doubt, due to the amount of effort directed at,
as well as to the differences in the ease of study of, the different
organisms.

The papers are reproduced here in a more or less taxonomic
order as this seemed the most logical classification that we could
imagine. We have attempted to standardise the presentation but it
is inevitable that considerable differences in style will persist.

The meeting was attended by over 60 delegates and proved to be a lively and enjoyable occasion. We hope that this record of the meeting will prove as stimulating, not least, in encouraging the use of data rather than its mere accumulation and in the wider application of approaches which are commonly used for some organisms but as yet not applied to others.

ACKNOWLEDGEMENTS

We thank the governing bodies of the three organisations concerned for approving the idea of a joint meeting and for encouraging their members to participate.

We also thank all the authors for meekly being persuaded to take part and for responding to our pleas for their manuscripts more or less on time.

We also express our appreciation to members of the technical staff of the Zoology Department, University of Hull for their help in running the meeting including R. Wheeler-Osman who improved many of the figures. Particular thanks are due to Christine Ware who not only helped enormously in running the meeting but who also took on the task of preparing the camera ready copy.

We also appreciate the encouragement, patience and help of Dr. Robert Andrews and Mr. John Matzka of the Plenum Publishing Company.

CONTENTS

ATTACHMENT TO SUSPENDED SOLIDS AS A STRATEGY OF

ESTUARINE BACTERIA

R. Goulder, E.J. Bent and A.C. Boak

Department of Plant Biology
The University
Hull, HU6 7RX

INTRODUCTION

Bacteria in sea water are frequently attached to suspended solids. These bacteria probably have several trophic roles. As well as the obvious one of heterotrophic degradation of organic matter, they are implicated in the incorporation of dissolved organic compounds into particulate matter (Paerl, 1974) and they might serve as a food source to grazing zooplankton (e.g. Heinle and Flemer, 1975; Heinle et al., 1977; Lenz, 1977a). It is sometimes asserted that most bacteria in sea water are attached to suspended solids (Wood, 1953; Seki and Kennedy, 1969; Seki, 1970, 1972). Direct microscopic counts, however, do not show that attached bacteria are, as a general rule, more abundant than free bacteria; e.g. (1) Wiebe and Pomeroy (1972) found that >80% of bacteria were free at open-ocean stations in the Antarctic, although attached bacteria were more abundant at N. American coastal and estuarine sites; (2) Taga and Matsuda (1974) found that free bacteria usually exceeded attached bacteria in both oceanic and coastal waters in the Pacific; (3) Zimmermann (1977) found that attached bacteria in the Kiel Bight (Baltic Sea) made up only 3-8% of total bacteria.

In estuarine waters it is probable that there is marked between-site variation in the relative importance of attached bacteria. In Fig. 1, attached bacteria (as % of total) in samples from estuaries in N.E. England, is plotted against suspended-solids concentration. The samples were of surface water collected (1975-77) mainly from the middle Humber at Hull (c.36 km upstream of Spurn Head) and from the outer Humber at Spurn. A few samples were also taken from the Tyne, Wear, Tees and Esk. For full details of sampling sites and times see Goulder (1977) and (for Spurn samples) Goulder et al. (1980).

1

Counts of attached and free bacteria were made using epifluorescence
microscopy after acridine-orange staining and concentration on black
0.22μm pore-size Millipore membrane filters (Jones and Simon, 1975).
In the case of attached bacteria it was assumed that bacteria on the
lower, hidden, surfaces of particles equalled those counted on the
upper surfaces, hence counts were doubled. Details of procedure and
microscope used are given in Goulder (1976). Suspended-solids
concentrations were determined by drying (overnight at 100°C) and
reweighing, after filtration through Whatman GF/C glass-fibre filters.

Figure 1 demonstrates between-site variation in the relative
importance of attached bacteria and indicates that attached bacteria
are most important at sites with high concentration of suspended
solids. In the Humber at Hull, suspended solids were generally
>100 mg/l and >80% of bacteria were attached; at Spurn, suspended
solids were usually <30 mg/l and <30% of bacteria were attached.
Results from the Tyne, Wear and Esk samples were similar to those
from Spurn, those from the Tees were intermediate between Hull and
Spurn. Also included in Fig. 1 are results based on mean values
(taken from Zimmermann, 1977 and Lenz, 1977b) at 2 m depth at four
sites in the Kiel Bight. Here, the proportion of attached bacteria
(<4%) and suspended-solids concentration (<5 mg/l) were both much
lower than at the sites in N.E. England.

THE SITUATION IN THE HUMBER ESTUARY AT HULL

The observation that most planktonic bacteria in the estuary at
Hull are attached to suspended solids led to further work on suspended
solids and their attached bacteria at this site.

Figure 2 shows size categories of particles in the twelve Hull
samples used in Fig. 1. These results were obtained by direct microscopic measurements (bright-field illumination, 1000 × magnification)
of 100 randomly-selected particles. The most frequent size category,
in all samples,was 5-9 μm; on average 47% of particles were in this
category while only 6% were >24 μm. Most particles (mean = 86%)
were aggregates of inorganic material and fibrous or amorphous organic
matter; only a few consisted of isolated mineral grains or fragments
of organic material. Weight loss on ignition (0.5 h at 400°C) of
solids from later Hull samples (1978-80, see below) indicated that
c. 10% of suspended solids was organic matter.

Examination of suspended solids from Hull under epifluorescence
microscopy (1600 × magnification) revealed that the proportion of
particle surface colonized by bacteria was quite small. Particles
usually had only a few individual bacteria or small colonies attached
to them, and were frequently observed without attached bacteria.
Similar observations were made on oceanic and estuarine suspended
solids by Wiebe and Pomeroy (1972) and Zimmermann (1977).

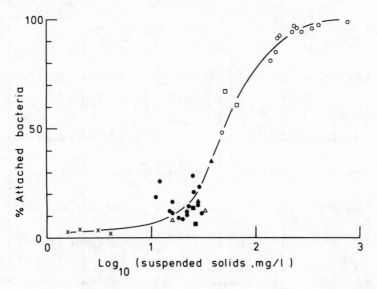

Figure 1. Plot of attached bacteria (as % of total bacteria) against suspended-solids concentration (log scale) in samples from estuaries in N.E. England. Samples from the Humber at Hull, July 1976 – January 1977 (**o**), and Spurn, June 1977 (**•**). Also from the Tyne at S. Shields, December 1975 (**△**); Wear at Sunderland, November 1975 (**▲**); Tees at Middlesborough, July 1975 (**□**); and Esk at Whitby, September 1975 (**■**). Also included are points based on mean values from the Kiel Bight (**×**).

Bacteria in estuaries are important because they bring about heterotrophic degradation of organic matter. It is interesting, therefore, to cite a study in which assessment was made of the relative contribution of both attached and free bacteria to heterotrophic activity (Bent and Goulder, in press). In this work V_{max} for glucose mineralization (i.e. the potential mineralization rate per unit volume at non-limiting substrate concentration) was determined at 10°C, using the Harrison, Wright and Morita (1971) modification of the method of Wright and Hobbie (1966), and served as an indicator of heterotrophic activity. The water samples were collected at high tide at monthly intervals, from January 1978 to January 1980, from a jetty off Albert Dock, Hull and V_{max} was obtained for whole samples and for samples with suspended solids (but not free bacteria) removed by centrifugation. The centrifuged samples provided V_{max} for free bacteria while the difference between V_{max} of whole samples and V_{max} of centrifuged samples gave V_{max} of attached bacteria.

Values of V_{max} for attached bacteria showed marked seasonal periodicity, being high during the autumn to spring period (maximum 5.7 µg $1^{-1}h^{-1}$ in November 1979) and low in summer (minimum 0.01 µg

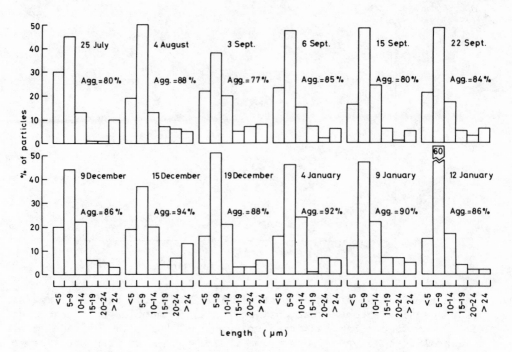

Figure 2.   Size categories of suspended-solid particles in samples
from the Humber at Hull,   July 1976 to January 1977.   The % of
particles which were aggregates is indicated.

$1^{-1}h^{-1}$ in June 1978). V_{max} of free bacteria (range 0.03 - 1.3 µg
$1^{-1}h^{-1}$) did not show this periodicity. V_{max} of attached bacteria
generally exceeded that of free bacteria.   This was most pronounced
in winter, while in summer there were exceptions on occasions when
very low values for attached bacteria were recorded.

Densities of attached bacteria and free bacteria were also
determined, using epifluorescence microscopy.   The density of attached
bacteria showed a seasonal periodicity similar to V_{max} of attached
bacteria (maximum 40.7×10^6/ml in January 1978, minimum 2.8×10^6/ml
in August 1978).   Free bacteria (range $0.8 - 7.0 \times 10^6$/ml) did not
show this periodicity and remained comparatively low throughout the
year.   The density of attached bacteria usually exceeded that of free
bacteria, especially during the autumn to spring period.

The seasonal periodicity of heterotrophic activity of attached
bacteria (as indicated by V_{max}) may therefore be explained by
variation in the density of attached bacteria.   Similarly, the fact
that activity of attached bacteria generally exceeds that of free
bacteria is explained by the greater density of attached bacteria.

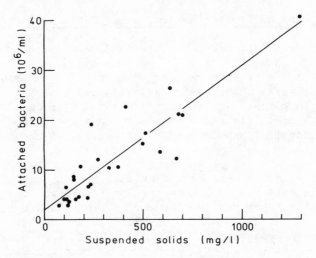

Figure 3. Plot of density of attached bacteria against suspended-solids concentration in samples from the Humber at Hull, January 1978 to January 1980.

Concentrations of suspended solids were also determined; these were high in winter and low in summer (maximum 1290 mg/1 in January 1978, minimum 74 mg/1 in August 1978). This seasonal pattern in the Humber depends on the interaction of several physical variables; e.g. temperature, freshwater flow, tidal range, wind conditions (Anon., 1970). In Fig. 3 the density of attached bacteria is plotted against suspended-solids concentration. There is some scatter, but the linear relationship indicates that, on the whole, the extent of colonization (i.e. the number of bacteria per unit weight of solids) was fairly constant. It follows that the seasonal periodicity in density of attached bacteria (and also heterotrophic activity) is probably primarily a result of variation in suspended-solids concentration.

POSSIBLE ADVANTAGES OF THE ATTACHED HABIT

The demonstration that most bacteria in the water column in the Humber at Hull are attached to suspended solids, and that attached bacteria are responsible for most heterotrophic activity, led to consideration of some of the possible advantages, to a bacterium in the middle estuary, of being attached to a particle rather than being free in the water. These were: (1) that attached bacteria may develop greater heterotrophic potential per cell; (2) that attached bacteria may have a nutritional advantage because of adsorption of dissolved organic compounds onto suspended solids; (3) that attached bacteria may be shielded from grazing zooplankton.

1.   Development of Greater Heterotrophic Potential per Cell

Attached bacteria might be able to develop greater average
heterotrophic potential per cell, perhaps because they are larger
than free bacteria or taxonomically different. Zimmermann (1977)
found that bacteria within detritus particles in the Kiel Bight
were larger than free bacteria, and Taga and Matsuda (1974) found
differences, in Pacific sea water, between attached and free bacteria
in their biochemical properties and proportion of chromogenic
individuals.

To investigate the situation at Hull, the average heterotrophic
potential per cell for both attached and free bacteria, in the samples
taken off Albert Dock between January 1978 and January 1980, was
calculated by Bent and Goulder (in press). The measure used was $V_{max}/$
bacterium for glucose mineralization which was obtained by dividing
V_{max} by the density of bacteria. The results are given in Table 1.
The values are approximate, because they are derived from two
variables both subject to error, but some conclusions can be made.
(1) Both attached and free bacteria showed considerable variation
in $V_{max}/$bacterium (the ranges were <0.1 - 2.9 $\times$ 10^{-10}µg/h for attached
bacteria, and <0.1 - 7.7 $\times$ 10^{-10}µg/h for free bacteria). (2) There
was no significant difference between the two sets of values ($P>0.05$,
Mann-Whitney U-test); the mean value of $V_{max}/$bacterium was actually
greater for free bacteria (Table 1) but this was due to a few
atypically high values for free bacteria in winter 1979-80.

It follows that, so far as $V_{max}/$bacterium for glucose mineralization serves as an indicator, there is no evidence that individual
attached bacteria at Hull have a greater heterotrophic potential
than free bacteria. It should be noted, however, that $V_{max}/$
bacterium has its limitations because it is a measure of potential,
not actual, average glucose-mineralization rate per cell. Also,
glucose is only one of many organic substrates utilized in nature;
bacteria might respond differently to substrates other than glucose.

2.   Nutritional Advantage because of Adsorption of Dissolved Organic Matter

Dissolved organic compounds, present in low concentrations in
the water, might be concentrated at particle surfaces by physico-
chemical adsorption and hence become more available to attached
bacteria. Many authors have suggested that this is an important
advantage of the attached habit (e.g. ZoBell, 1943, 1946; Jannasch,
1958; Jannasch and Pritchard, 1972; Sorokin, 1978). There are,
however, two separate processes by which dissolved organic compounds
can be incorporated into particulate matter. These are, (1) physico-
chemical adsorption, and (2) direct biological uptake into the cells
of attached micro-organisms. If the second process predominates
then the nutritional advantage of the attached habit is less obvious

Table 1. Vmax per Bacterium for Glucose Mineralization (10^{-10} µg/h) by Attached Bacteria and Free Bacteria. Samples from the Humber at Hull, January 1978 to January 1980.

Date (1978)	Attached bacteria	Free bacteria	Date (1979-80)	Attached bacteria	Free bacteria
30 January	-	0.1	4 January	0.3	0.6
23 February	0.2	0.1	26 February	0.9	0.3
22 March	0.6	0.2	26 March	2.9	0.8
18 April	0.2	0.6	20 April	0.5	0.4
15 May	<0.1	<0.1	15 May	0.4	0.6
12 June	0.5	0.2	11 June	0.7	0.2
6 July	0.3	0.3	10 July	0.3	0.6
1 August	0.3	0.3	1 August	1.3	3.0
29 August	1.0	2.0	30 August	0.4	2.6
21 September	1.5	0.4	25 September	1.1	1.9
17 October	0.3	0.9	19 October	2.2	4.2
13 November	0.6	0.4	27 November	-	2.0
7 December	-	0.4	17 December	-	7.7
			22 January	2.3	4.8
			Mean	0.8	1.3

(-) Indicates no data available.

because dissolved organic matter is equally available for direct
uptake by both free and attached bacteria.

With dissolved organic compounds at artificially high concentration, physico-chemical incorporation into marine particulate matter
can be demonstrated. For example, Rosenfeld (1979) shook a suspension
of clay-sediment particles in sea water with mixtures of three amino-
acids, each at c. 50-290 mg/l, and obtained adsorption which he judged
to be non-biological. Under natural conditions, however, incorporation of dissolved organic compounds into particulate matter might be
principally microbiological. Paerl (1974), for example, incubated
natural marine and freshwater samples with added tritiated glucose
and acetate, at 0.001 μgl^{-1}, and showed by autoradiography that the
substrate incorporated into particulate matter was taken up directly
by attached bacteria.

Some experiments were, therefore, carried out to determine
whether incorporation of dissolved organic substances into particulate
material, in the Humber at Hull, is chiefly microbiological or physico-
chemical.

Estuary water was incubated, with ^{14}C-labelled glucose or amino-
acid mixture, under three treatments, and mineralization to $^{14}CO_2$
and ^{14}C incorporation into particulate matter were measured. The
treatments were, (1) no treatment, (2) with 1 mg/l of copper (added
as $CuSO_4$) to bring about partial inhibition of bacteria, and (3)
sterilized by autoclaving.

To measure mineralization, 25 ml sub-samples (3 replicates and
a control under each treatment) with added ^{14}C glucose (c. 3 $\mu g/l$
and c. 4 $\mu Ci/l$) or amino-acid mixture (c. 1 μg amino-acid C/l and
c. 4 $\mu Ci/l$) were shaken for 2 h at $10^{\circ}C$ in 120-ml serum bottles.
The amino-acid mixture used consisted of fourteen ^{14}C-labelled amino-
acids mixed in approximately the same proportion as in typical algal-
protein hydrolysate (Radiochemical Centre, Amersham, CFB 104). The
serum bottles were sealed with a gas-tight rubber serum cap fitted
with a glass rod and cup which held a concertina-folded square
(c. 4 × 4 cm) of Whatman No. 1 filter paper above the surface of the
sub-sample (illustrated by Fry and Humphrey, 1978). The incubations
were stopped by injection, through the serum caps, of 2 ml of 2.5M
H_2SO_4 into the sub-samples; this killed bacteria and released the
$^{14}CO_2$ produced during incubation. Next, 0.25 ml of 2-phenylethylamine
was injected onto the folded-paper squares; this was allowed to
absorb $^{14}CO_2$ overnight. The papers were then transferred to vials
and $^{14}CO_2$ was determined by liquid scintillation counting.

To measure incorporation into suspended solids, an identical
set of incubations was set up, but in serum bottles closed with
simple screw caps. After 2 h incubation, a 2.0 ml sub-sample from
each bottle was membrane filtered (Nuclepore, 3 μm pore size, 2.5 cm

diam.). The filters were washed-through with 5 ml of filtered estuary water, solubilized (Amersham NCS tissue solubilizer) and ^{14}C
incorporated was determined by liquid scintillation counting.

The results obtained using a surface-water sample (suspended
solids = 801 mg/l) taken at high tide from Corporation Pier, Hull
on 1 February 1979, are given in Table 2. Values are percentages
of total isotope supplied which was either mineralized or incorporated into suspended solids. With both substrates, treatment with
copper caused partial inhibition of both mineralization and incorporation. In the sterile samples, total inhibition of mineralization
was accompanied by virtually complete inhibition of incorporation.
These results suggest that incorporation of dissolved organic compounds into suspended solids at Hull is principally microbiological.
If incorporation were physico-chemical then inhibition of bacterial
respiration would not have been accompanied by parallel inhibition
of adsorption.

Table 2. Mineralization to ^{14}CO$_2$, and Incorporation onto Suspended
          Solids, of ^{14}C-labelled Dissolved Glucose and Amino-Acid
          Mixture during Incubation under Three Treatments. Sample
          from the Humber at Hull, 1 February 1979.

Treatment	Glucose		Amino-acids	
	Mineraliz.	Incorp.	Mineraliz.	Incorp.
Untreated	7.8	28.5	9.1	27.1
+ 1 mg Cu/l	2.3	7.5	0.5	5.2
Autoclaved	0.0	0.3	0.0	0.4

Values are percentage of total isotope supplied which was
mineralized or incorporated.

The experiment was repeated twice, using samples taken 15
February (suspended solids = 819 mg/l) and 29 March 1979 (1307 mg/l),
and similar results were obtained. The data from all three experiments were used to plot percentage of added substrate incorporated
against percentage mineralized (Fig. 4). There was a close linear
relationship between the two processes. The regression line indicates
that when there was no bacterial activity (i.e. at zero mineralization), mean incorporation, over 2 h, equalled 1.2% of available
substrate. For untreated samples, incorporation ranged from 27-56%
of substrate. Hence perhaps c. 2-4.5% of total incorporation was
non-microbiological. This value compares quite well with the <5%
quoted by Paerl (1974) for non-microbiological adsorption onto freshwater particulate material.

Since physico-chemical adsorption of dissolved organic compounds

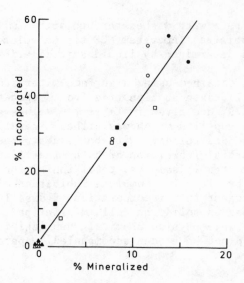

Figure 4. Plot of % of added ^{14}C-labelled substrate which was incorporated into suspended solids against % mineralized to $^{14}CO_2$
during 2 h incubation with estuary water from Hull. Symbols are as
follows: (○) glucose (●) amino acids, no treatment; (□) glucose
(■) amino acids, 1 mg Cu/l; (△) glucose (▲) amino-acids, auto-
claved.

onto suspended solids is apparently relatively unimportant, there
is no evidence to support the suggestion that adsorption gives
attached bacteria a nutritional advantage over free bacteria in the
middle section of the Humber estuary.

3.   Protection from Grazing Zooplankton

Attachment to suspended solids might provide protection from
grazing zooplankton. At first sight it seems likely that attached
bacteria, being on particles apparently of suitable size for ingestion by zooplankton (Fig. 2), might be more liable to suffer from
grazing than free bacteria which, because of their small size
(<1 μm), might be grazed inefficiently. It is possible, however,
that attached bacteria at Hull are not heavily grazed because they
make up, on average, only about 0.06% of suspended solids. This
figure was derived from a mean value of 2.9×10^7 attached bacteria/
mg of suspended solids (Fig. 3) and a mean dry weight per bacterium
of 2.0×10^{-14} g. The mean dry weight was obtained by measuring
c. 150 bacteria, in a Hull sample taken December 1979, using epifluorescence microscopy, calculating mean volume and assuming a
density of 1 g/cm^3 and a dry weight to wet weight ratio of 0.2
(Jones, 1979). It follows that a grazing animal needing say 20% of

its body weight per day for maintenance would, if it relied on
attached bacteria, need to ingest >330 times its body weight per
day. The zooplankton may avoid the ingestion of such a large bulk
of largely inorganic material by either utilizing the non-living
organic component of the suspended solids, or by selecting their
food, perhaps microalgae, on the basis of particle size or quality.

To consider the susceptibility of bacteria to grazing, investigations were made, at intervals throughout 1979, on the composition
of the zooplankton community and on the extent of grazing on attached
and free bacteria.

Vertical hauls (from 8 m to the surface) were made at slack
high-water, from the jetty off Albert Dock, using a Plymouth-pattern
net (46 cm diam., 150 × 150 μm mesh size). Catches from three hauls
were combined and animals in sub-samples were identified and counted.
Much the most important species in the zooplankton was the brackish-
water calanoid copepod Eurytemora hirundoides Nordquist (Table 3).
Also recorded were other calanoid copepods, cyclopoid and harpacticoid
copepods, and the mysid Neomysis integer (Leach) but their densities
were so low that only grazing by Eurytemora is considered here.

Table 3. The Composition of the Zooplankton Community in the Humber
          at Hull, 1979.

Date	Eurytemora hirundoides (adults + copepodites)	Other calanoid copepods	Cyclopoid copepods	Harpacticoid copepods	Neomysis integer
17 January	204	0	0	4	4
16 March	1110	0	0	22	4
14 April	1459	0	10	24	1
14 May	1353	0	0	0	0.5
12 June	2775	0	10	20	111
12 August	91	4	0	3	17
8 September	395	3	0	12	1.5
7 October	185	3	0	1	1
5 November	382	6	0	8	2
4 December	1101	0	0	13	0
Mean	906	1.6	2.0	10.7	14.2

Values are animals/m^3.

Grazing was measured in the field, under as near-natural
conditions as possible, using recently collected zooplankton and
^{14}C-labelled natural bacteria as food source. To prepare labelled
bacteria, estuary water was incubated with ^{14}C glucose (c. 30 μg/l
and c. 50 μCi/l) for 8 h at estuary temperature. Labelled attached
bacteria, together with their suspended solids, were then separated
by low-speed centrifuging (15 min. at 400 R.C.F.), washed, and
resuspended at original density in sterile filtered estuary water.

Labelled free bacteria in the initial supernatant were concentrated
by high-speed centrifuging (15 min. at 3000 R.C.F.) and were also
washed and resuspended. Immediately before grazing determinations,
animals in a suitable volume were concentrated into 50 ml of estuary
water (without exposure to air or entrapment against a net) and were
transferred to a 250-ml cylindrical plastic chamber together with
50 ml of the estuary water which contained the labelled free or
attached bacteria. Three replicates, and two controls with anaes-
thetized or dead animals, were run with each food type. Ten minutes
grazing (less than the gut-retention time) was allowed, in subdued
light at estuary temperature, with occasional gentle swirling to
prevent settling of suspended solids. Grazing was stopped by anaes-
thetizing the animals with CO_2, to prevent reflex evacuation of gut
contents (Burns and Rigler, 1967), and fixing with formalin. Animals
were then sorted and counted, solubilized (NCS solubilizer), and
their radioactivity was determined by liquid scintillation counting.
Mean ^{14}C ingested per individual Eurytemora (for adults and cope-
podites combined) was then calculated. Mean grazing rate per individual equalled the mean ^{14}C ingested per individual divided by the
total isotope supplied in food bacteria times the total number of
bacteria introduced to the grazing chamber divided by the duration
of grazing.

Eurytemora population grazing rates on attached and free
bacteria (as number of bacteria eaten per m^3 per day, and as % of
bacterial population eaten/day) were obtained by combining individual grazing rates with values of Eurytemora population density and
density of attached and free bacteria in surface water.

The results are given in Table 4. Values of number and percentage grazed, for both attached and free bacteria, varied considerably but some conclusions can be made. (1) The number of bacteria
eaten was always far too low to maintain the Eurytemora population -
presumably the animals mainly eat food other than bacteria. (2)
The percentage of both attached and free bacteria removed by grazing
was always too small to influence significantly the bacterial
populations (maximum values were only c. 0.01%/day for attached
bacteria and c. 0.06%/day for free bacteria).

Because the Eurytemora population grazing rates on both attached
and free bacteria were very low, and because the number and percentage of attached bacteria grazed was not significantly different
from the number and percentage of free bacteria grazed (in both
cases P>0.05, Mann-Whitney U-test), there is no evidence to support
the suggestion that differential grazing, at Hull, favours attached
bacteria over free bacteria.

CONCLUSION

Although most bacteria at Hull are attached to suspended solids,

Table 4. Grazing Rates, for the whole <u>Eurytemora</u> <u>hirundoides</u>
Population, on Attached Bacteria and Free Bacteria in the
Humber at Hull, 1979.

Date	Number of bacteria eaten per day $(10^8/m^3)$		% of bacterial population eaten per day x 1000	
	Attached	Free	Attached	Free
17 January	0.1	0.8	0.1	8.6
16 March	1.8	2.1	0.6	7.7
14 April	10.9	12.7	11.3	36.5
14 May	10.5	3.4	1.9	1.8
12 June	9.8	2.5	6.6	17.1
12 August	1.1	0.1	0.4	0.6
8 September	5.2	0.4	8.2	2.3
7 October	1.0	0.3	0.9	3.2
5 November	0.3	0.4	0.7	1.8
4 December	13.1	6.0	7.0	55.9
Mean	5.4	2.9	3.8	13.6

and most heterotrophic activity is associated with suspended solids,
we have not identified the advantages, to an individual bacterium,
of being attached to a particle.

The following are further possible factors which may be favourable to attached bacteria but which we have not investigated. (1)
Particulate organic matter, which makes up c. 10% of suspended solids,
may be utilized by attached bacteria but is not available to free
bacteria. (2) Suspended solids at Hull settle from time to time.
Attached bacteria may, therefore, spend considerable time in the
sediments, hence they are perhaps less likely to be washed out to
sea than free bacteria. (3) Attachment to suspended solids may
provide protection from dissolved toxic pollutants (e.g. heavy
metals).

Further investigation is clearly required to explain the success
of attached bacteria in estuarine situations; consideration of
metabolic requirements and taxonomy might be particularly rewarding.

ACKNOWLEDGEMENTS

Part of this work was supported by a NERC research grant.
E.J. Bent held a NERC CASE studentship, in co-operation with the
Yorkshire Water Authority, for which he is grateful. We thank
Pamela Jackman for her technical assistance.

REFERENCES

Anon. (1970). Silt movement in the Humber Estuary. British Trans-
    port Docks Board, Research Station Report No. R221, 1-8.
Bent, E.J. and Goulder, R. (in press). Planktonic bacteria in the
    Humber Estuary; seasonal variation in population density and
    heterotrophic activity. Mar. Biol.
Burns, C.W. and Rigler, F.H. (1967). Comparison of filtering rates
    of Daphnia rosea in lake water and in suspensions of yeast.
    Limnol. Oceanogr. 12: 492-502.
Fry, J.C. and Humphrey, N.C.B. (1978). Techniques for the study of
    bacteria epiphytic on aquatic macrophytes. In: Techniques for
    the Study of Mixed Populations (ed. by D.W. Lovelock and R.
    Davies). Soc. appl. Bact. tech. Ser. 11: 1-29. Academic Press,
    London.
Goulder, R. (1976). Relationships between suspended solids and
    standing crops and activities of bacteria in an estuary during
    a neap-spring-neap tidal cycle. Oecologia 24: 83-90.
Goulder, R. (1977). Attached and free bacteria in an estuary with
    abundant suspended solids. J. appl. Bact. 43: 399-405.
Goulder, R., Blanchard, A.S., Sanderson, P.L. and Wright, B. (1980).
    Relationships between heterotrophic bacteria and pollution in
    an industrialized estuary. Wat. Res. 14: 591-601.
Harrison, M.J., Wright, R.T. and Morita, R.Y. (1971). Method for
    measuring mineralization in lake sediments. Appl. Microbiol.
    21: 698-702.
Heinle, D.R. and Flemer, D.A. (1975). Carbon requirements of a
    population of the estuarine copepod Eurytemora affinis. Mar.
    Biol. 31: 235-247.
Heinle, D.R., Harris, R.P., Ustach, J.F. and Flemer, D.A. (1977).
    Detritus as food for estuarine copepods. Mar. Biol. 40: 341-
    353.
Jannasch, H.W. (1958). Studies on planktonic bacteria by means of
    a direct membrane filter method. J. gen. Microbiol. 18: 609-
    620.
Jannasch, H.W. (1958). Studies on planktonic bacteria by means of
    particulate matter in the activity of aquatic microorganisms.
    Mem. Inst. Ital. Idrobiol. 29: 1-112.
Jones, J.G. (1979). A guide to methods for estimating microbial
    numbers and biomass in fresh water. Scient. Publs Freshwat.
    biol. Ass. 39: 317-329.
Jones, J.G. and Simon, B.M. (1975). An investigation of errors in
    direct counts of aquatic bacteria by epifluorescence microscopy,
    with reference to a new method for dyeing membrane filters.
    J. appl. Bact. 39: 317-329.
Lenz, J. (1977a). On detritus as a food source for pelagic filter-
    feeders. Mar. Biol. 41: 39-48.
Lenz, J. (1977b). Seston and its main components. In: Microbial
    Ecology of a Brackish Water Environment (ed. by G. Rheinheimer).
    pp. 37-60. Springer-Verlag, Berlin.

Paerl, H.W. (1974). Bacterial uptake of dissolved organic matter in relation to detrital aggregation in marine and freshwater systems. Limnol. Oceanogr. 19: 966-972.

Rosenfeld, J.K. (1979). Amino acid diagenesis and adsorption in nearshore anoxic sediments. Limnol. Oceanogr. 24: 1014-1021.

Seki, H. (1970). Microbial biomass on particulate organic matter in seawater of the euphotic zone. Appl. Microbiol. 19: 960-962.

Seki, H. (1972). The role of microorganisms in the marine food chain with reference to organic aggregate. Mem. Ist. Ital. Idrobiol. 29: Suppl. 245-259.

Seki, H. and Kennedy, O.D. (1969). Marine bacteria and other heterotrophs as food for zooplankton in the Straight of Georgia during the winter. J. Fish. Res. Bd Can. 26: 3165-3173.

Sorokin, Y.I. (1978). Decomposition of organic matter and nutrient regeneration. In: Marine Ecology 4, Dynamics (ed. by O. Kinne). pp. 501-616. Wiley, New York.

Taga, N. and Matsuda, O. (1974). Bacterial populations attached to plankton and detritus in seawater. In: Effect of the Ocean Environment on Microbial Activities (ed. by R.R. Colwell and R.Y. Morita). pp. 433-448. University Park Press, Baltimore.

Wiebe, W.J. and Pomeroy, L.R. (1972). Microorganisms and their association with aggregates and detritus in the sea: a microscopic study. Mem. Ist. Ital. Idrobiol. 29: Suppl. 325-352.

Wood, E.J.F. (1953). Heterotrophic bacteria in marine environments of eastern Australia. Aust. J. mar. freshwat. Res. 4: 160-200.

Wright, R.T. and Hobbie, J.E. (1966). Use of glucose and acetate by bacteria and algae in aquatic ecosystems. Ecology 47: 447-464.

Zimmermann, R. (1977). Estimation of bacterial number and biomass by epifluorescence microscopy and scanning electron miscroscopy. In: Microbial Ecology of a Brackish Water Environment (ed. by G. Rheinheimer). pp. 103-120. Springer-Verlag, Berlin.

ZoBell, C.E. (1943). The effect of solid surfaces upon bacterial activity. J. Bact. 46, 39-56.

ZoBell, C.E. (1946). Marine Microbiology. Chronica Botanica Co., Waltham, Massachusetts.

GROWTH AND SURVIVAL OF ESTUARINE MICROALGAE

I.R. Joint

Natural Environment Research Council
Institute for Marine Environmental Research
Prospect Place, The Hoe, Plymouth, PL1 3DH

INTRODUCTION

Estuarine microalgae have to adapt to environmental conditions not usually experienced by marine phytoplankton; the salinity variation within estuaries is considerable and, in addition, many estuaries have very turbid water with shallow euphotic zones. Intertidal estuarine microalgae must adapt to extremes of salinity, to desiccation and high light intensities and must resist removal by tidal scour.

The purpose of this review is to consider some of the mechanisms by which estuarine microalgae survive. Both benthic and planktonic microalgae will be considered and their response to such environmental factors as salinity and light intensity will be discussed.

RESPONSE OF BENTHIC MICROALGAE TO LIGHT

Benthic algae growing on intertidal areas are exposed to considerable variations in light intensity; in addition to diurnal changes in solar radiation, the light regime varies from day to day because the period of tidal exposure changes with a tidal periodicity. Intertidal algae have to accommodate a much wider range of light conditions than phytoplankton and they do, in fact, appear to be better adapted to both high and low light intensities.

Benthic microalgae are much less sensitive to high light intensities than are phytoplankton. Taylor (1964) found very little photoinhibition at "full sunlight" of 75 cal cm^{-2} h^{-1} (ca 2100 μE m^{-2} s^{-1}; for ease of comparison, all light measurements quoted in this paper will also be converted to μEm m^{-2} s^{-1}) but also that this phytobenthic

community would photosynthesise efficiently at low light levels.
Photosynthesis was saturated at 12 cal cm^{-2} h^{-1} (ca 300 µE m^{-2} s^{-1})
and cells receiving only 1% incident solar radiation (0.75 cal
cm^{-2} h^{-1}, ca 20 µE m^{-2} s^{-1}) were still able to fix carbon at 35% of
their maximum rate. By measuring the light penetration into the
sand, Taylor (1964) calculated that diatoms existing as deep as 3 mm
in the sediment would be above their compensation level and that those
at 2 mm depth would be capable of photosynthesis at greater than 90%
of the maximum.

The absence of photoinhibition was also recorded by Colijn and
Buurt (1975) who found that photosynthesis of a natural population
was saturated by 5000 lux (ca 100 µE m^{-2} s^{-1}) at 6°C and by 10,000
to 12,000 lux (ca 200-240 µE m^{-2} s^{-1}) at 12°C; Cadée and Hegeman
(1974) did not apparently saturate photosynthesis with 1 cal cm^{-1}
mm^{-1} (ca 1670 µE m^{-2} s^{-1}) at a temperature of 16°C. In contrast,
Morris and Glover (1974) studied three species of phytoplankton in
culture and found no effect of the growth temperature on the saturat-
ing light intensity. However, care is needed in extrapolating from
culture to field conditions; this problem is illustrated by the
results of Colijn and Buurt (1975) who found that, in contrast to
experiments with natural populations, a culture of the benthic diatom
Amphiprora alata did show photoinhibition at light intensities greater
than 10,000 lux (ca 200 µE m^{-2} s^{-1}) at 12°C. However, the authors
point out that the culture was grown at a light intensity of 3,000
lux (ca 60 µE m^{-2} s^{-1}) and the cells would not be adapted to high
light conditions.

The effect of light on the growth of benthic diatoms was studied
in culture by Admiraal (1977a) who found that the length of the
illumination period had a significant effect on the growth rate at
different temperatures; at 12°C Amphiprora c.f. paludosa grew fastest
with an 8h photoperiod but at 20°C, the highest growth rate was
obtained with a 16h illumination period; i.e. the optimum day length
for Amphiprora paludosa is shorter at low temperatures than at higher
temperatures. Admiraal's (1977a) data with cultures suggested that
the response of benthic microalgae to high light intensity was not
significantly different to that of phytoplankton; this observation
is at variance with the data of Taylor (1964), Colijn and Buurt
(1975) and Cadée and Hegeman (1974) which showed the absence of photoinhibition in natural populations and again emphasises the problem
of extrapolating from laboratory culture to field conditions.

It is not known what mechanism benthic diatoms use to maintain
maximum photosynthesis at very high light intensities. Absence of
photoinhibition has also been reported for one species of the green
macroalgae, Codium fragile by Arnold and Murray (1980) but the
tolerance to high light was attributed to the thick, optically dense
morphology of the plant. Benthic diatoms could, of course, migrate
into the sediment at times of high light intensity and so reduce the

light reaching the cells. This would be feasible in muddy sediments
where movement over very small distances would drastically reduce
the light intensity, but it is less likely to be a mechanism in
sandy sediments where the required distances would be much greater.
Haardt and Nielsen (1980) measured the light attenuation by sand and
mud and reported attenuation coefficients of 1.9 mm^{-1} for sand and
12.2 mm^{-1} for mud, corresponding to a 1% light intensity level at
depths of 2.4 mm for sand and 0.38 mm for mud. Taylor (1964) also
measured the 1% light intensity level at 3 mm depth in sand. Therefore, migration in unlikely to be the mechanism of high light
tolerance in sand because the distances are too great. In any case,
Taylor's (1964) and Colijn and Buurt's (1975) measurements were made
on natural populations separated from the sediment and so shading of
the algae by the sediment must be discounted as the mechanism of
high light tolerance. The biochemical mechanism resulting in the
absence of photoinhibition in benthic microalgae is worthy of further
study.

The response of benthic microalgae to light intensity is affected
by a number of factors other than temperature, such as salinity and
desiccation. McIntire and Wulff (1969), using a laboratory model
ecosystem, found most growth of microalgae at high light intensities
when the community was not subject to desiccation. The community
structure was affected by desiccation and diversity was reduced in
the upper littoral areas which were exposed to long periods of high
light and exposure. These results were obtained in the summer but in
a subsequent, winter study, Wulff and McIntire (1972) found much less
effect of light intensity and desiccation on the distribution of the
diatoms although, of course, the biomass of diatoms was considerably
reduced.

VERTICAL MIGRATION OF BENTHIC DIATOMS

Unlike phytoplankton cells that are dependent on water movement
to bring them into a favourable light regime, many species of benthic
microalgae can control the amount of light received by migrating into
and out of the sediment. It is a common observation that the surface
of an intertidal mudflat is intensely coloured brown or green at low
water but that this colour is absent immediately after exposure of
the mudflat. The increase in colour on the sediment surface is the
result of benthic microalgae migrating out of the sediment to optimum
light conditions. This vertical migration has been utilized in the
quantitative sampling method of Eaton and Moss (1966) which reduces
the light received by the algae by placing lens tissue on the sediment
surface; the algae respond by migrating into the lens tissue to a
more favourable light intensity and the lens tissue is removed after
several hours with the quantitative recovery of motile algae.

Early workers postulated that the rhythm of vertical migration
was tidal, controlled by a "biological clock" and was not a response

to light (for a brief history, c.f. Perkins, 1960). However, Perkins
(1960) concluded that diatoms did not exhibit a tidal rhythm but they
migrated with a diurnal rhythm that depended only on the intensity
of light. Hopkins (1963) studied the vertical distribution of a
diatom community and found it confined to the top 2 mm of mud; the
movement by the diatoms was confined to 1 mm and was stimulated by
physical shock which Hopkins (1963) suggested would be similar to
that caused by the wave action of a receding tide. However, the
primary stimulus for upward migration in the mud appeared to be light
intensity.

An extensive study of the motile, benthic alga, Euglena obtusa
in the field and in the laboratory was made by Palmer and Round
(1965). E. obutsa, a major constituent of the mud banks of the river
Avon at Bristol, reached densities greater than 10^5 cells cm^{-2} and
caused distinct green colouration of the mud surface at low tide.
Migration in and out of the mud was found to have a tidal rhythm but
cells would not emerge from the mud in the absence of light and cells
already on the surface could be made to migrate into the sediment by
darkening the surface. However, this was more than a direct response
to light because, when removed to the laboratory and kept under constant illumination, the migrating rhythm persisted with a diurnal,
and not a tidal, amplitude. Palmer and Round speculated that the
primarily phototactic response of E. obtusa was modified by the very
high turbidity of the water, exposing the cells to periods of darkness
at high water which imposed a tidal rhythm on the algal migration.

Round and Palmer (1966) studied the vertical migration of diatoms from the same site and found that they also migrated into the
sediment before the tide reached them. The rhythm of migration
could be maintained for several days in cells transferred to the
laboratory and they concluded that the migration was under the control
of a "biological clock" and was not a phototactic response. The
degree of rhythm varied from species to species and a change in
temperature resulted in a shift in the "biological clock". However,
in all cases, the rhythm in the laboratory was diurnal and the only
tidal effect obtained in the laboratory was when tidal flooding
occurred during daylight hours; this resulted in a downward migration
which carried over to subsequent days.

In a laboratory study with the benthic diatom, Surirella gemma,
Hopkins (1966) also found a diurnal rhythm controlled by daylength
and by the moisture of the mud; wetting of the algal community in
darkness resulted in diatoms migrating into the mud. However, not
all diatom species in a tidal flat show vertical migration; Round
(1979) studied the vertical migration of two benthic diatoms isolated
from Barnstable Harbor, Mass. Hantzschia virgata var intermedia
descended into the sand on a flood tide but Tropidoneis lepidoptera
isolated from the same sand flat, did not move with a tidal periodicity. Round (1979) speculated that this behaviour could result in

large numbers of T. lepidoptera being swept away as the sandflat
was flooded.

It is not clear what the evolutionary significance is of vertical
migration by benthic microalgae. Obviously it is a response which
enables an algal cell to receive optimum light but it is debatable
whether or not this behaviour has evolved as a mechanism of escaping
the scouring effect of a flood tide.

SEDIMENT STABILIZATION BY BENTHIC DIATOMS

The characteristics of intertidal sediments are modified by the
mucilage produced by benthic diatoms making the sediments more stable
and less susceptible to resuspension; algae which occur in such
intertidal areas will, therefore, be less likely to be removed by
tidal scour. However, as Round (1979) pointed out, some species
which produce mucilage also descend into the mud on the flood tide
and would appear not to benefit from the mucilage. The mucilage
excreted by benthic diatoms is not confined to the surface, euphotic,
layer of the sediment; Frankel and Mead (1973) found that the inter-
stices of a pebbly sand were filled with mucilage to a depth of at
least 7 cm. In a more detailed study of six species of benthic diatoms in culture, Holland et al. (1974) found very effective stabilization of fine sediments by those species which secreted large amounts
of mucilage, resulting in decreased resuspension of both algal cells
and sediment. The production of mucilage is necessary for the
locomotion of diatoms (Gordon and Drum, 1970) but Holland et al.
(1974) speculated that mucilage producation is more than a consequence
of locomotion and that it has a selective advantage for intertidal
microalgae enabling them to remain in optimum light conditions without the danger of being removed by the tidal scour.

BURIAL AND DARK SURVIVAL OF BENTHIC MICROALGAE

Benthic and microalgae inevitably become buried in the sediment
and significant concentrations of chlorophyll a have been reported at
depths of up to 15 cm in sediment by many workers (e.g. Taylor and
Gebelein, 1966; Steele and Baird, 1968; Fenchel and Straarup, 1971;
Cadée and Hegeman, 1974; Joint, 1978). This chlorophyll a is
present far below the euphotic zone of the sediment, presumably buried
as a result of animal activity (cf. Stockner and Lund, 1970). All
of these studies assumed that chlorophyll a was an indication of viable algae, because of the rapid degradation of chlorophyll a to
phaeopigments in dead cells. As a test of this assumption, a comparison was made of the usual method of measuring extracted chlorophyll a,
with a microscopic examination of sediment; viable cells were assumed
to be those that showed the characteristic red fluorescence of
chlorophyll when examined with a blue light fluorescence microscope.
Figure 1 shows that there is a good agreement between the two methods
near the surface of the sediment but, at depth, the pigment extraction

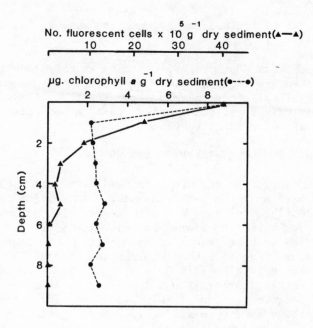

Figure 1. Comparison of extracted chlorophyll a with microscopic counts of algal cells which have the characteristic chlorophyll fluorescence. Microscopic counts were made with a blue light epifluorescence microscope and the pigment extraction procedure was that used by Joint (1978) who also described the study site in the Lynher estuary, S.W. England.

over estimates the number of viable algae buried in the sediment. Even so, there are considerable numbers of apparently viable cells buried in sediment.

Laboratory studies have shown that it is possible for an algal cell to retain viability for long periods in the dark. Smayda and Mitchell-Innes (1974) tested nine planktonic diatoms and found that seven retained viability for 90 days in darkness. Antia (1976) found that benthic algae survived better than phytoplankton in darkness; most benthic strains survived 12 months and some retained viability after 3 years of darkness. If the prolonged viability also occurs in algae buried in the sediment, and the results shown in Fig. 1 suggest that this is so, sediment turnover may return cells to the surface euphotic zone, even after burial for many months, and growth may resume.

EFFECT OF SALINITY ON BENTHIC MICROALGAE

Estuarine benthic microalgae are exposed to much larger changes in salinity than phytoplankton because they cannot move with the

water mass; for example, in the Lynher estuary, over a tidal cycle,
benthic algae can be exposed to salinity changes greater than 20%
in the water and similar rapid changes in salinity are common in
other estuaries. Salinity changes pose problems to the cells in
maintaining cell turgor and the passage of nutrients and water across
the cell membrane, but estuarine benthic diatoms appear to be
particularly tolerant to such changes (Zimmermann and Steudle, 1971;
Wulff and McIntire, 1972; McIntire and Reimer, 1974). Admiraal
(1977a) found that photosynthesis of unialgal cultures and mixed
isolates was largely unaffected within the salinity range 4 to 60‰;
in a subsequent laboratory study, Admiraal (1977b) ascribed most of
the changes in diatom density and species composition to grazing by
ciliates and not to changes in salinity. However, Moore and McIntire
(1977) suggested that <u>mean</u> salinity accounted for most of the
differences in diatom distribution along the Yaquina estuary, Oregon.
In a study of a heavily organically polluted estuarine mudflat,
Admiraal and Peletier (1980) found that some diatom species inhabited
only a restricted part of the possible salinity range determined in
a laboratory; however, in such a nutrient rich area, a number of
factors other than salinity will influence community structure.

EFFECT OF SALINITY ON ESTUARINE PHYTOPLANKTON

     Surprisingly, there have been relatively few studies of the
effect of salinity per se on estuarine phytoplankton even though
phytoplankton experience larger salinity changes in estuaries than
in any other marine environment. Mandelli et al. (1970) found a
"different phytoplankton cycle" in tidal estuaries of Long Island,
N.Y., than in the adjacent coastal water with a preponderance of
green flagellates and dinoflagellates in the estuaries and an
alternation of diatoms with dinoflagellates in the coastal water.
Flemer (1970) found the largest concentration of phytoplankton in
the low saline areas of Chesapeake Bay, which he attributed partly
to the nutrient supply from the river although components of marine
origin must also be involved. The effect of salinity on the species
composition of phytoflagellate blooms in New York Bay was studied
by Mahoney and McLaughlin (1979) who found that the salinity optima
of three typical bloom species in culture was the same as the
salinity of the waters in which the algae were found. They concluded,
however, that salinity tolerance was not an important factor in
bloom development and that maximum growth rate of the individuals
controlled the species composition of the bloom.

     Vosjan and Siezen (1968) studied the effect of salinity on the
photosynthetic rate of two algae in culture; the marine alga,
<u>Chlamydomonas uva-maris</u> was very tolerant to salinity change with
very little change in photosynthesis between 10‰ and 100‰ . In
contrast, the freshwater alga <u>Scenedesmus obliquus</u> was more sensitive
and photosynthesis decreased gradually with increasing salinity,
until photosynthesis equalled respiration at about S = 30‰ . The

effects of ionic composition of the medium have been extensively
studied with algal cultures (Craigie, 1969; Borowitzka and Brown,
1974; Kirst, 1977; Kauss, 1978) and it is clear that, although
organic cell constituents are involved in the regulation of osmotic
pressure, the major contribution in algal cells is the cation, K^+.
However, there is a need to expand these laboratory studies to the
field, to confirm that these mechanisms apply to phytoplankton within estuaries.

SURVIVAL OF FRESHWATER PHYTOPLANKTON

Morris et al. (1978) have recently raised the question of the
fate of freshwater algae entering an estuary. These workers observed
a sharp drop in dissolved oxygen at very low salinities (the fresh-
water-seawater interphase, FSI) in the Tamar estuary, S.W. England;
the oxygen sag occurred at times of high phytoplankton biomass in
the river and coincided with a peak in dissolved organic carbon.
Morris et al., hypothesised that the mechanism of the oxygen sag
was a mass mortality of freshwater halophobic phytoplankton as a
result of the osmotic changes occurring at this low salinity region,
with concomitant release of organic material and its assimilation
by the heterotrophic microbes, resulting in the utilization of
dissolved oxygen and producing the observed oxygen sag. The salinity
at which the dissolved organic carbon peak and oxygen sag occurred
was less than 1‰. The fate of freshwater algae entering estuaries
has not received much study and, in view of the obvious relationship
between algal growth in the river and the oxygen sag at the FSI, it
is important to know what role freshwater phytoplankton play in
estuaries.

Blanc et al. (1969), measuring physical and ecological parameters at the mouth of the Rhône, reported a great abundance of
freshwater phytoplankton in the dilution zone (salinity of ca 8 to
26‰) but most of the cells were "dead or almost dead". In contrast,
Foester (1973) made a very detailed study of the fate of freshwater
algae in an estuarine system and concluded that a large number of
freshwater algae did survive passage from the river. Using neutral
red dye to determine the pH of the cell contents, he could distinguish between living and dead or senescent cells and found that many
freshwater species in the estuary were viable. Samples of estuarine
water incubated with freshwater media developed cultures of freshwater algae, confirming the validity of the neutral red technique.
The increased salinity resulted in morphological changes in many
species. At salinities of up to 10‰, Chlamydomonas angulosa and
Chlorella vulgaris swelled and the chloroplasts became misshapen,
Oocystis parvula formed a thickened cell wall and Navicula sp.
became larger; in contrast, Euglena sanguiena appeared to be des-
troyed by salinities greater than 10‰. Some of the morphological
changes were extreme and Foester speculated whether some cells
would be unrecognizable as algae and might be identified as organic

detritus in a microscopic examination of an estuarine sample. However, the conclusion of his study was that many freshwater algae are capable of surviving transport from freshwater to seawater.

There have been few attempts to culture freshwater algae in seawater but Wetherell (1961) found that many freshwater algae grew well in enriched seawater. Figure 2 shows data from Wetherell (1961) recalculated and plotted as the biomass obtained after culture in seawater media of different salinities as a percentage of the control growth in freshwater medium. The amount of growth varied between species, but, of the 13 algae tested by Wetherell, only one was unable to grow at salinities greater than 7‰ and 8 grew at salinities of more than 21‰. No attempt was made by Wetherell to precondition the algae to increased salinity and it is possible that better growth might have resulted if the cells had experienced a gradual change in salinity, as would occur in passage down an estuary.

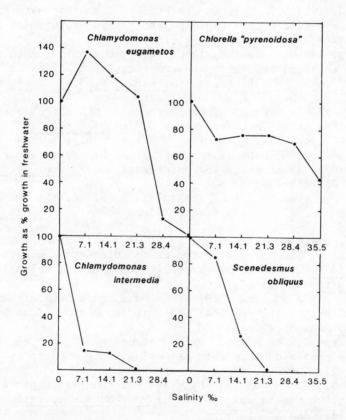

Figure 2. The growth of four freshwater phytoplankton in different salinity seawater, expressed as a percentage of the biomass obtained in freshwater media. Data recalculated from Wetherall (1961).

These data highlight our ignorance of the role of freshwater
algae in the ecology of estuaries and it may even be appropriate to
question the long held view that the freshwater flora is funda-
mentally different from the marine flora. Estuaries clearly offer
a unique environment in which to study these important questions on
the physiology and ecology of microalgae.

ACKNOWLEDGEMENT

This work forms part of the Estuarine Ecology programme of the
Institute for Marine Environmental Research, a component of the
Natural Environment Research Council.

REFERENCES

Admiraal, W. 1977(a). Influence of light and temperature on the
    growth rate of estuarine benthic diatoms in culture. Mar. Biol.
    39: 1-9.
Admiraal, W. 1977(b). Salinity tolerance of benthic estuarine dia-
    toms as tested with a rapid polarographic measurement of photo-
    synthesis. Mar. Biol. 39: 11-18.
Admiraal, W. and H. Peletier, 1980. Distribution of benthic species
    on an estuarine mud flat and experimental analysis of the
    selective effect of stress. J. exp. mar. Biol. Ecol. 46:
    157-175.
Antia, N.J. 1976. Effects of temperature on the darkness survival
    of marine microplanktonic algae. Microbial Ecology 3: 41-54.
Arnold, K.E. and Murray, S.N. 1980. Relationships between irradi-
    ance and photosynthesis for marine benthic green algae
    (Chlorophyta) of differing morphologies. J. exp. mar. Biol.
    Ecol. 43: 183-192.
Blanc, F., M. Leveau and K.H. Szekielda. 1969. Effets eutrophique
    au débouché d'un grand fleuve (Grand Rhône). Mar. Biol. 3:
    233-242.
Borowitzka, L.J. and A.D. Brown. 1974. The salt relation of marine
    and halophilic species of the unicellular green alga,
    Dunaliella; the role of glycerol as compatible solute. Arch.
    Microbiol. 96: 37-52.
Cadée, G.C. and J. Hegeman. 1974. Primary production of the benthic
    microflora living on tidal flats in the Dutch Wadden Sea. Neth.
    J. Sea Res. 8: 260-291.
Colijn, F. and G. van Buurt. 1975. Influence of light and temperature
    on the photosynthesis rate of marine benthic diatoms. Mar.
    Biol. 31: 209-214.
Craigie, J.S. 1969. Some salinity-induced changes in growth, pigments
    and cyclohexanetetrol content of Monochrysis lutheri. J. Fish
    Res. Bd. Canada 26: 2959-2967.
Eaton, J.W. and B. Moss. 1966. The estimation of numbers and pigment
    content in epipelic algal populations. Limnol. Oceanogr. 11:
    584-595.

Fenchel, T. and B.J. Straarup. 1971. Vertical distribution of photo-
     synthesis pigments and the penetration of light in marine
     sediments. Oikos 22: 172-182.
Flemer, D.A. 1970. Primary production in the Chesapeake Bay. Ches.
     Sci. 11: 117-129.
Foester, J.W. 1973. The fate of freshwater algae entering an estuary.
     In: Estuarine Microbial Ecology, ed. L.H. Stevenson and R.R.
     Colwell, University of South Carolina Press, pp. 387-420.
Frankel, L. and D.J. Mead. 1973. Mucilaginous matrix of some
     estuarine sands in Connecticut. J. sedim. Petrol. 43: 1090-1095.
Gordon, R. and R.W. Drum. 1970. A capillary mechanism for diatom
     gliding locomotion. Proc. natn. Acad. Sci. U.S.A. 67: 338-344.
Haardt, H. and G.E. Nielsen. 1980. Attenuation measurements of mono-
     chromatic light in marine sediments. Oceanol. Acta. 3: 333-338.
Holland, A.F., R.G. Zingmark and J.M. Dean. 1974. Quantitative
     evidence concerning the stabilization of sediments by marine
     benthic diatoms. Mar. Biol. 27: 191-196.
Hopkins, J.T. 1963. A study of the diatom of the Ouse estuary,
     Sussex. 1. The movement of the mudflat diatoms in response to
     some chemical and physical changes. J. mar. biol. Assoc. U.K.
     43: 653-663.
Hopkins, J.T. 1966. The role of water in the behaviour of an
     estuarine mudflat diatom. J. mar. biol. Ass. U.K. 46: 617-626.
Joint, I.R. 1978. Microbial production of an estuarine mudflat.
     Est. coast. Mar. Sci. 7: 185-195.
Kauss, H. 1978. Osmotic regulation in algae. In: Progress in phyto-
     chemistry. 5: 1-27. Ed. Reinhold, L. et al., Pergamon.
Kirst, G.O. 1977. Ion composition of unicellular marine and fresh-
     water algae, with special reference to Platymonas subcordiformis,
     cultivated in media with different osmotic strengths. Oecologia
     (Berl.) 28: 177-189.
Mahoney, J.B. and J.J.A. McLaughlin. 1979. Salinity influence on
     the ecology of phytoflagellate blooms in lower New York Bay
     and adjacent waters. J. exp. mar. Biol. Ecol. 37: 213-223.
Mandelli, E.F., P.R. Burkholder, T.E. Doheny, and R. Brody. 1970.
     Studies of primary productivity in coastal waters of southern
     Long Island, New York. Mar. Biol. 7: 153-160.
McIntire, C.D. and B.L. Wulff. 1969. A laboratory method for the
     study of marine benthic diatoms. Limnol. Oceanogr. 14: 667-678.
McIntire, C.D. and C.W. Reimer. 1974. Some marine and brackish
     water Achnanthes from Yaquina estuary, Oregon (U.S.A.) Botanica
     mar. 17: 164-175.
Moore, W.W. and C.D. McIntire. 1977. Spatial and seasonal distribu-
     tion of littoral diatoms in Yaquina estuary, Oregon (U.S.A.).
     Botanica mar. 20: 99-109.
Morris, A.W., R.F.C. Mantoura, A.J. Bale and R.J.M. Howland. 1978.
     Very low salinity regions of estuaries; important sites for
     chemical and biological reactions. Nature 274: 678-680.
Morris, I. and H.E. Glover. 1974. Questions on the mechanism of
     temperature adaptation in marine phytoplankton. Mar. Biol.

24: 147–154.

Palmer, J.D. and F.E. Round. 1965. Persistent vertical migration rhythms in benthic microflora. I. The effect of light and temperature on the rhythmic behaviour of Euglena obtusa. J. mar. biol. Assoc. U.K. 45: 567–582.

Perkins, E.J. 1960. The diurnal rhythm of the littoral diatoms of the river Eden estuary, Fife. J. Ecol. 48: 725–728.

Round, F.E. and J.D. Palmer. 1966. Persistent, vertical migration rhythms in benthic microflora. II. Field and laboratory studies on diatoms from the banks of the river Avon. J. mar. biol. Ass. U.K. 46: 191–214.

Round, F.E. 1979. Occurrence and rhythmic behaviour of Tropidoneis lepidoptera in the epipelon of Barnstable Harbor, Massachusetts, U.S.A. Mar. Biol. 54: 215–217.

Smayda, T.J. and B. Mitchell-Innes. 1974. Dark survival of auto-trophic, planktonic marine diatoms. Mar. Biol. 25: 195–202.

Steele, J.H. and I.E. Baird. 1968. Production ecology of a sandy beach. Limnol. Oceanogr. 13: 14–25.

Stockner, J.G. and J.W.G. Lund 1970. Live algae in postglacial lake desposits. Limnol. Oceanogr. 15: 41–58.

Taylor, W.R. 1964. Light and photosynthesis in intertidal benthic diatoms. Helgoländer wiss. Meersunter. 10: 29–37.

Taylor, W.R. and C.D. Gebelein. 1966. Plant pigments and light penetration in intertidal sediments. Helgoländer wiss. Meersunter. 13: 229–237.

Vosjan, J.H. and R.J. Siezen. Relation between primary production and salinity of algal cultures. Neth. J. Sea Res. 4: 11–20.

Wetherell, D.F. 1961. Culture of freshwater algae in enriched natural seawater. Physiologia Pl. 14: 1–6.

Wulff, B.L. and C.B. McIntire. 1972. Laboratory studies of assemblages of attached estuarine diatoms. Limnol. Oceanogr. 17, 200–214.

Zimmermann, U. and E. Steudle. 1971. Effects of potassium concen-tration and osmotic pressure of seawater on the cell-turgor pressure of Chaetomorpha linum. Mar. Biol. 11: 132–137.

SURVIVAL STRATEGIES OF ATTACHED ALGAE IN ESTUARIES

Martin Wilkinson

Department of Brewing and Biological Science
Heriot-Watt University
Chambers Street
Edinburgh, EH1 1HX

INTRODUCTION

There is little published evidence concerning the mechanisms of survival of attached algae in estuaries. At the cellular level, studies on osmotic and ionic regulation in algae have largely been directed at using the algae as convenient tools for the investigation of physiological processes rather than at using physiological differences between algae to explain differences in their distributions. At the whole plant level, while many workers have investigated environmental tolerances of algae in batch culture, results have sometimes been contradictory and difficult to relate to observed field distributions. This has resulted in some cases from inadequacies in the design of such experiments. A comprehensive review of the literature concerning the distribution of estuarine algae in relation to environmental factors has been given by Wilkinson (1980). It is not proposed to repeat this review but instead to highlight several promising approaches, worthy of further investigation, to explain the survival of estuarine algae. These include the simultaneous adoption of more than one strategy, the existence of subspecific variants in widely distributed species and the growth patterns of the algae themselves.

ATTACHED ALGAE OF ESTUARIES

It is necessary, firstly, to describe the distribution patterns shown by attached estuarine algae. These can be summarised as follows:

1. Colonisation almost wholly by marine species with freshwater ones abundant only in the uppermost reaches of the estuary.

29

2.   Reduction in species number going upstream due to selective
      attenuation of red, then of brown algae. Green algae and blue-
      green algae do not necessarily become more numerous, in terms
      of species, going upstream but they constitute a much greater
      proportion of the algal community.

3.   Colonisation of the mid-reaches by brackish water algae such
      as Fucus ceranoides and certain Vaucheria spp.

There is a trend from the larger more foliose algae of the open
coast to the filamentous, mat-forming algae of structural simplicity,
fast growth rate and short-life span of the inner estuary.

SIMULTANEOUS ADOPTION OF MULTIPLE STRATEGIES

      Because of the reduction in species number and, therefore,
possibly in interspecific competition, the increased physical rigour
of the estuarine environment, and the apparent reduction of macrofaunal grazers in the upper reaches of some estuaries, there has
been a belief that the primary determinants of estuarine algal
distribution are abiotic. This contrasts with the open coast where
algal zonation is primarily determined by biotic factors (Chapman,
1973). The only laboratory study on competition in estuarine macroalgae would support this. Russell and Fielding (1974), using a
modification of the de Wit replacement series technique for studying
competition in crop plants, showed that three species of filamentous
algae, Erythrotrichia carnea, Ectocarpus siliculosus and Ulothrix
flacca, had strong competitive interactions in sea water media but
became less competitive in low salinity media. Unfortunately,
while these species can colonise lower reaches of estuaries, they
only rarely occur in mid and upper reaches and so this evidence
is of limited use. In an unpublished survey of the algal flora of
72 estuaries around Britain, Wilkinson has found relatively few
colonising species which, however, tend to occur in most estuaries
sampled. Apparent differences in algal vegetation between the
estuaries were more a matter of which particular species were
abundant than of which ones were present. The same situation occurs
with the well-marked vertical zonation observed in some small
stratified estuaries (Wilkinson, Fuller et al., 1981) where,
distinct horizontal bands, each dominated by different species, can
be seen. When percentage frequency of occurrence of each species
is measured on vertical transects the major zone-forming species are
found to occur over all or most of the vertical extent of the shore
(Fig. 1). These two facts suggest that the algae concerned are
widely tolerant of physical conditions and that competition may be
less important than on the open coast. As yet there is no information
on the effect of meiofaunal grazers and this could be an important
biotic factor in the upper reaches of estuaries. Clearly, the
relative importance of biotic and physical influences is unresolved.
It is, however, wise to expect an alga to adopt more than one strategy
as the following example shows.

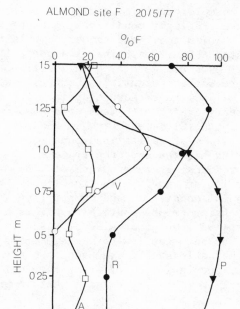

Figure 1. Percentage frequency of four different dominant species of algae on a vertical transect on the river wall of the Almond estuary near Edinburgh. To the naked eye a three zone shore was visible, dominated successively by Rhizoclonium riparium (R), Vaucheria sp. (V) and Phormidium spp. (P). Audouinella spp. (A) were locally dominant at all levels where intense shade was provided. Note that the dominants have a wider distribution than their visually apparent zone.

Melosira nummuloides is a filamentous, chain-forming, centric diatom which dominates the lower shore in the upper reaches of the Clyde estuary and some other highly polluted estuaries. It is present only as a trace form in most estuaries studied in Britain (Wilkinson, unpublished data). Wilkinson et al. (1976) pointed out that it was dominant in the part of the estuary where the principal macrofaunal grazers, littorinids, were absent. The distributions of the two were mutually exclusive and contiguous, and transplant experiments showed that M.nummuloides could survive with Littorina spp. in the estuary only when protected from grazing. Hence grazing was believed to determine the areas of dominance of M.nummuloides. This cannot be the complete explanation for this dominance because other species, principally filamentous green and blue-green algae, flourish in other estuaries in the absence of severe macrofaunal grazing. Rendall and Wilkinson (in preparation)

have shown that M.nummuloides is particularly tolerant to a wide
range of temperature and light conditions, factors which are
particularly variable in the Clyde estuary. Wilkinson et al.,
(1976) also suggested that it might be at a competitive advantage
in an organically - enriched estuary with low light penetration if
it had a wide heterotrophic potential. Corrigan and McLean (1979)
and McLean et al. (1980) have shown that isolates of M.nummuloides
from the Clyde estuary are able to take up a number of organic carbon
sources, enriched in the Clyde estuary, in the dark, and that uptake for some is maximum at salinities found in the diatom's area
of dominance. Having had the opportunity to occur abundantly in
the absence of macrofaunal grazers, M.nummuloides may have two
strategies fitting it for survival in the Clyde. These are wide
tolerance to stressful conditions and competitive advantage due to
possible accessory heterotrophic nutrition. This allies well with
the ideas of Grime (1977) on strategies of vascular plants; ideas
which have not so far been applied to algae. Grime suggested that
under different combinations of disturbance and stress found in
different habitats, species with different survival strategies would
be dominant as shown in Table I. He said that in practice few

Table I   Suggested Basis for the Evolution of Three Strategies in
          Vascular Plants

Intensity of disturbance	Intensity of stress	
	low	high
low	competitive strategy	stress-tolerant strategy
high	ruderal strategy	no viable strategy

from Grime, J.P. 1977. American Naturalist, 111,
1169-1194.

habitats would be extreme and in most there would be a mixture of
strategies depending on the mixture of stress and disturbance
prevailing. For the short-lived estuarine algae the habitat is more
likely to be stressful than disturbed and correspondingly a mixture
of stress-tolerant and competitive strategies could be expected, as
found in the case of Melosira above. Grime traced the preponderance
of different strategies through seral successions and found that the

competitive-stress tolerant combination was characteristic of climax
communities of potentially high productivity such as woodlands and
also of areas of decreasing productivity. Both these descriptions
could apply to the Melosira community dominant as a climax vegetation
in a sewage-polluted estuary. It would be interesting to explore
further the application of Grime's ideas to algal vegetation as
this is only a rather superficial example presented here.

Stress Tolerance

The only stress factor to have been extensively investigated
for multicellular algae is that of salinity. Tolerance has been
assessed in many ways by different workers including long-term
growth measurements and short-term photosynthesis/respiration
measurements. It is very difficult to compare the results of
different workers, as pointed out by Yarish et al. (1979b), because
of differences in experimental technique and pretreatment of the
plants. Wilkinson, Roe et al. (1981) measured the survival,
the rates of photosynthesis and respiration at different salinities,
after a range of pretreatments, of a wide range of estuarine species.
In almost all cases a wide tolerance to salinity was found, with
little difference existing in photosynthesis and respiration rates
for estuarine algae at different salinities.

Unfortunately, this kind of experiment, which is the most often
attempted, can be criticised on a major point. Salinity fluctuates
rapidly in the field in most estuarine situations and den Hartog
(1967) has pointed out that it is the fluctuation of salinity rather
than the absolute value which determines tolerance. He showed that
the same species penetrated the stable salinities of the Baltic
Sea to lower absolute salinites than those to which they penetrated
in the fluctuating salinities of West European estuaries. As yet,
there are almost no published data on growth of estuarine algae in
simulated fluctuating salinities and this must be ultimately done
before salinity can be properly assessed as a stress factor.
Wilkinson, Fuller et al. (1981) have suggested that the well
marked vertical zonation pattern of small stratified estuaries might
be correlated with the five aspects of the combined salinity and
emersion factors shown in Table II. This is illustrated by Fig. 2.

Burrows (1964) cultured Fucus serratus and F.ceranoides under
different salinity regimes. The former is normally restricted to
the open coast and the latter to estuaries. In culture, both
tolerated brackish salinities in constant salinity conditions but
when a daily salinity fluctuation was introduced, only F.ceranoides
survived. This underlines the need for salinity fluctuation in
experimental investigations. Burrows also circumvented a second
problem in the interpretation of salinity tolerance experiments;
that often only one age or life-history stage is used. While she
showed the adult plants of the coastal F.serratus tolerated a

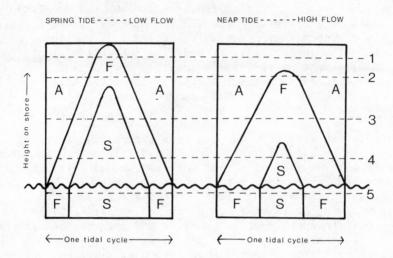

Figure 2. Hypothetical tide curves incorporating marked salinity
stratification for two contrasting sets of conditions in a small
stratified estuary. A, atmospheric emersion; S, submersion in the
seawater layer; F, submersion in the freshwater layer. 1-5
represent five different vertical position on the shore. Note the
variation in salinity environment between sites 1-5:
1. Freshwater only. Submerged on spring tides only.
2. Freshwater only. Submerged on all tides.
3. Fresh ⟶ salt ⟶ fresh on spring tides; freshwater only
     on neap tides.
4. Fresh ⟶ salt ⟶ fresh on all tides.
5. Fresh ⟶ salt ⟶ fresh. Always submerged.
Note also the relative lengths of fresh and salt submergence on
different tides and river flow conditions.

Table II Main physical variants with increasing height in
           an estuarine intertidal zone

1. Relative submersion/emersion time
2. Frequency of submersion
3. Relative saltwater/freshwater times during a
     submersion period
4. Frequency of saltwater/freshwater submersion
5. Maximum and minimum salinities and their rate
     of change

constant low salinity, she also showed that this was probably not realised in the field, F.serratus would be unlikely to establish populations at such low salinities as F.ceranoides. This was because low salinity tolerance was developed almost immediately in zygotes of the latter but did not develop for some hours in those of the former species.

Stress-tolerance is not merely an abiotic factor effect but can be elicited by the biotic factor of competition. Kindig and Littler (1980) have suggested four strategies that may be involved in increasing the competitive ability of Ulva around a sewage outfall in California. There is no reason why these could not be applied to an estuarine situation. The four strategies are given by the following hypotheses:

1. Acclimation hypothesis   – tolerant froms become acclimated to stressful conditions and hence survive i.e. phenotypic modification.

2. Adaptation hypothesis   – long-term exposure to stress has caused natural selection of genetically adapted forms.

3. Eurytopic hypothesis   – tolerant forms are inhibited but wide tolerance enables survival of higher stress than species with narrower tolerance.

4. Enhancement hypothesis   – growth rates of tolerant species are enhanced by some aspect of the stressful environment and hence confer competitive superiority.

There is insufficient evidence to evaluate all these hypotheses for estuarine algae but the recent spate of work on subspecific variation in estuarine algae provides evidence on the first two hypotheses.

Subspecific Variation

It is proposed that widely distributed algal species exist in a wide range of conditions because the species contains several variants each adapted to a different part of the distribution range. These variants could be either phenotypic modifications, i.e. acclimated physiological races, or genetically distinct ecotypes. For example, Geesink (1973) has shown that the freshwater red alga Bangia atropurpurea and the marine B.fuscopurpurea can be acclimated to tolerate each others' environment over a number of culture generations. On the other hand, Russell and Bolton (1975) showed that isolates of the brown alga Ectocarpus siliculosus from different salinity environments retained their differences in salinity tolerance over four years in culture through many generations,

under different salinites, indicating genetic separation of races.
Recently Yarish et al. (1979a) showed that there were genetic
differences between ecotypes of the red algae Caloglossa and
Bostrychia from different positions along an estuarine gradient
in the U.S.A. but that each genetically distinct ecotype also had
some capacity for phenotypic modification. Hence, a combination
of both approaches is active in this example. This situation is
further confused by some preliminary work reported by Wilkinson,
Roe et al. (1981). A coastal and an estuarine population of the
green alga Enteromorpha intestinalis showed a difference, in accord
with their habitat, in the salinity optimum for photosynthesis and
this was heightened by pretreatment of the plants for 2 weeks under
the salinity of the test. This suggests some kind of adaptive
difference between the populations. However, other populations of
the same species from estuarine and coastal environments showed a
broad tolerance with no particular salinity optimum. It, therefore,
seems that there is a possibility that different strategies may
prevail in different population of the same species.

The maintenance of genetic differences between populations
requires some degree of reproductive isolation. However, many
estuarine algae show a suppression of reproduction with decreased
salinity (see Wilkinson, in press for references) and this may
effect their potential for subspecific variation. The Forth
Estuary provides a further example. In the upper reaches there
is a population of Enteromorpha intestinalis found only in the
Summer. In Winter the inner limit is further downstream. The
Winter salinities at this population's shore level are almost
constantly freshwater and this seems to restrict its occurrence.
In the summer, the salinity rises with low freshwater flow, and
hence greater salt penetration, to about 6‰ S. This is a low
enough level to restrict reproduction in the marine E.intestinalis
(Wilkinson, in press) and this particular field population has not
been found in a reproductive condition. How then is it propagated?
It may arise from propagules coming in with the salt water in the
Summer. If this is so, the population is likely to be genetically
similar to those further downstream.

Some Other Features of Estuarine Algae

Just as reproduction may be inhibited under low salinities, so
also growth may be suspended. The open coast green alga Monostroma
oxyspermum penetrates stratified estuaries in Britain but grows
low on the shore when it will be in the salt wedge for part of the
tidal cycle. It seems to need higher salinities. In some estuaries
where it is abundant, however, e.g. the Clyde, long periods of high
freshwater flow mean that it can be in freshwater for several days.
Culture experiments carried out by Wilkinson (unpublished) show
that while it cannot grow in freshwater media, it can survive for
several weeks and resume growth on transfer back to high salinity

media. Rendall and Wilkinson (in preparation) have found the same
with Melosira nummuloides from the Clyde.

The life-form of estuarine attached algae is important. They
are generally those types characteristic of disturbed or stressed
areas on the open sea coast and characteristic in that habitat of
earlier stages in a seral succession. They are short-lived, fast-
growing algae of simple construction. They frequently form mats
in which one species gradually covers another until the whole mat
peels off and floats away. In the author's experience this can
happen at some prolific sites several times a year and observations
of denuded areas (Wilkinson, Fuller et al., 1981) show that recolonisation occurs within a few weeks. Hence, the algae may pass
through several generations in a year and so there is the possibility
of evolution occurring in a cyclic pattern over an annual seasonal
cycle. Indeed, preliminary unpublished results of the author
suggest that seasonal variation in photosynthetic response of the
widely-tolerant populations of E.intestinalis mentioned earlier
may be more adaptive than spatial variation along an estuary at any
one season. There may, therefore be an adaptation to seasonal
changes in estuarine conditions in some populations or species
while there are adaptations to local differences in others. The
pattern of subspecific variation in estuarine algae is, therefore,
yet more complicated and would appear to encompass not one, but
several survival strategies.

Research Needs

It is surprisingly difficult to review a field in which there
is little published information. This review has deliberately
concentrated on only a few lines on which moderate amounts of data
are available but even then has had to reply heavily on hypothesis
and on unpublished results. It is, therefore, suggested that in
order to understand factors affecting estuarine algal distributions
and to evaluate their potential role in environmental assessment,
work is urgently needed along the following lines:

1.   The effects of fluctuating environmental factors, particularly
     salinity.
2.   The role of heterotrophy in estuarine macroalgae.
3.   The role of competition between estuarine algae.
4.   The changed patterns of grazing on algae in upper reaches of
     estuaries particularly grazing by meiofauna.
5.   The genetic structure of estuarine algal populations.

ACKNOWLEDGEMENTS

The author is indebted to the Carnegie Trust for the Universities
of Scotland for financial assistance with some of his estuarine field
work reported here; and for helpful discussion, to his students

Lesley MacLeod, Ian Fuller, Nigel Rudd, Fiona Roe and particularly
to David Rendall and David Mills.

REFERENCES

Burrows, E.M. 1964. Ecological experiments with species of Fucus.
     Proc. int. Seaweed Symp., 4: 166-170.
Chapman, A.R.O. 1973. A critique of prevailing attitudes towards
     the control of seaweed zonation on the sea-shore. Botanica
     mar., 16: 80-82.
Corrigan, J. and McLean, R.O. 1979. Nutrition of an epilithic
     estuarine diatom. Br. phycol. J., 14: 122.
Geesink, R. 1973. Experimental investigations on marine and fresh-
     water Bangia (Rhodophyta) from the Netherlands. J. exp. mar.
     Biol. Ecol., 11: 239-247.
Grime, J.P. 1977. Evidence for the existence of three primary
     strategies in plants and its relevance to ecological and
     evolutionary theory. American Naturalist, 111: 1169-1194.
den Hartog, C. 1967. Brackish water as an environment for algae.
     Blumea, 15: 31-43.
Kindig, A.C. and Littler, M.M. 1980. Growth and primary productivity
     of marine macrophytes exposed to domestic sewage effluents.
     Marine Environ. Res., 3: 81-100.
McLean, R.O., Corrigan, J.C. and Webster, J. 1980. Further observa-
     tions on Melosira nummuloides from the Clyde estuary. Br.
     phycol. J., 15: 196.
Russell, G. and Bolton, J.J. 1975. Euryhaline ecotypes of Ectocarpus
     siliculosus (Dillw.) Lyngb. Estuar. cstl. mar. Sci., 3: 91-94.
Russell, G. and Fielding, A.H. 1974. The competitive properties of
     marine algae in culture. J. Ecol., 62: 689-698.
Wilkinson, M. 1980. Estuarine benthic algae and their environment:
     A review. In: Systematics Association Special Volume no 17 (b)
     "The Shore Environment, Vol. 2: Ecosystems", edited by
     J.H. Price, D.E.G. Irvine and W.F. Farnham. Academic Press,
     London and New York.
Wilkinson, M., Fuller, I., Penny, J.W., Scanlan, C.M. and Roe, F.
     1981   Vertical zonation of intertidal algae in some small
     stratified estuaries. Phycologia. in press.
Wilkinson, M., Henderson, A.R. and Wilkinson, C. 1976. Distribution
     of attached algae in estuaries. Mar. Pollut. Bull., 7: 183-184.
Wilkinson, M., Roe, F., Taggart, P., Tollervey, A., Mackie, L. and
     MacLeod, L. 1981   Comparative salinity tolerances of
     freshwater, brackish and marine algae. Phycologia. in press.
Yarish, C., Edwards, P. and Casey, S. 1979a. A culture study of
     salinity responses in ecotypes of two estuarine red algae.
     J. Phycol., 15: 341-346.
Yarish, C., Edwards, P. and Casey, S. 1979b. Acclimation responses
     to salinity of three estuarine red algae from New Jersey.
     Marine Biology, 51: 289-294.

SURVIVAL STRATEGIES OF MEIOFAUNA

R.M. Warwick

Natural Environment Research Council
Institute for Marine Environmental Research
Prospect Place
The Hoe, Plymouth, PL1 3DH

INTRODUCTION

Mechanisms of survival fall into two broad categories. To survive in the short term animals must, of course, be adapted to the particular environments in which they live. However, other properties of animal populations can be regarded as "adaptations to the pattern of the environment in space and time" (Levins, 1968) involving flexibility of response to environmental factors, and it is these adaptations which are categorised as strategies. Efficiency and stability are the two features necessary for survival under natural selection (MacArthur, 1955) and it is the purpose of this paper to discuss certain aspects of the ecology of meiofauna populations and communities which ensure the fulfillment of these two criteria.

The repercussions of miniaturisation in the meiofauna are considerable, so much so that the strategies involved in maintaining efficiency and stability are entirely different from those of the macrofauna. I will consider these strategies under three headings: feeding, metabolism and reproduction/life history.

FEEDING STRATEGIES

Meiofauna species living together in the same sediment as the macrofauna are potentially in competition with them for the same primary food sources, particularly unicellular autotrophs and bacteria. However, the food gathering apparatus in meiofauna is relatively small in relation to the size of food particles, each particle usually being processed individually with a necessarily high degree of selectivity with respect to size, shape and quality (e.g. Lee et al., 1977; Tietjen and Lee, 1977), whereas the larger macrofauna have the ability to feed on a wide range of food-particle sizes in a relatively unspecialised manner (Calow, 1977). This can be regarded as just one aspect of the relationship between environmental grain, in the sense of MacArthur and Levins (1964), and the size of the organisms living in that environment. This relationship has important implications for community structure because, to quote Hutchinson (1959) "... small size, by permitting animals to become specialised to the conditions offered by small diversified elements of the environmental mosaic, clearly makes possible a degree of diversity quite unknown among groups of larger organisms" (see also Hutchinson and MacArthur, 1959).

MacArthur (1955) has shown that a given stability can either be achieved by a large number of species with specialised diets or by a smaller number of species with catholic diets and, at least in shallow water, we find that the species diversity of the meiofauna is almost invariably higher than that of the macrofauna, a fact which at once bewilders and frustrates newcomers to the field of meiofauna study. To quote MacArthur, "natural selection operates for maximum efficiency subject to a certain necessary stability". It seems that meiofauna species achieve efficiency by virtue of their restricted diet and because restricted diet per se lowers stability, this must be compensated for by a high species diversity. Indeed, meiofauna communities are usually remarkably stable (Warwick, 1980).

In the literature, the high diversity of meiofauna has often been attributed to factors other than food specialisation. Gray (1978), for example, invoked biological interactions (such as competition for resources and predator-prey interactions within the meiofauna) to explain the phenomenon. He based his reasoning on the fact that, in some examples which he studied, diversity was considerably higher than would be predicted from a neutral model (Caswell, 1976) which eliminated the influence of biological interactions on faunal diversity. Such interactions have also been suggested by other authors (e.g. McIntyre, 1971; Coull and Vernberg, 1975) as explaining high diversity.

However, a detailed analysis of the diversity (H') of the various faunal components of an intertidal mud flat in the Lynher estuary, Cornwall, U.K. (based on data in Warwick and Price (1975), Teare (1978), Warwick and Price (1979), Warwick et al., (1979) and other data in manuscript form) indicates that the diversity of each infaunal component is remarkably close to values predicted from the neutral model (Table 1); in this case there is no need to invoke competition or predation to explain observed diversities.
To calculate diversity, species abundance data from large samples averaged over a year have been reduced to 500 individuals using the rarefaction method of Sanders (1968), in the knowledge that

TABLE 1. Calculated values of diversity (H') of the faunal compo-
          nents of Lynher Estuary mud, and values predicted from a
          neutral model

	Meiofauna	Nematodes	Macrofauna
Calculated	2.992	2.813	1.212
Predicted	2.997	2.733	1.161

species numbers may, if anything, be slightly higher using this
method than they would be in a real sample of this size (Simberloff,
1972). This reduction was necessary in order that the numbers fell
within the range of H' values predicted from the neutral model given
in Table 2 (p.339) of Caswell (1976), interpolation between data
points of this table being made by multiple curvilinear regression.
Log_e was used for the calculation of both observed and predicted
values of H'.

Not only is the species diversity of the meiofauna at this
site much higher than that of the macrofauna, but it also appears
that the meiofauna are able to partition the total energy available
to them rather evenly amongst species, in comparison with the macrofauna. Fig. 1 shows the partitioning of annual production amongst
species of macrofauna (data from Warwick and Price, 1975) and
amongst the nematode component of the meiofauna (from Warwick,
1980). In common with other shallow-water communities studied, the
macrofauna production is dominated by one or a few species, whereas
with the nematodes production is partitioned remarkably evenly,
without dominance by a few species. We must, therefore, question
how the meiofauna maintain such a high diversity and how they partition available resources so equitably amongst species, and yet appear
to avoid competitive interactions. Temporal resource sharing can be
ruled out, since the species composition of the nematodes remains
quite stable seasonally (Warwick and Price, 1975). The environment
at this site can be regarded, at least superficially, as fine-
grained: the modal particle diameter is about 15 μm so that geometric
partitioning of interstitial voids as might occur in coarser sands
is not possible. The meiofauna are confined to the top few centimetres of sediment so that vertical stratification of species is
unlikely, even in response to the rather fine scale banding of
chemical and microbiological sediment characteristics described by
Anderson and Meadows (1978). Structuring of the environment by
macrofauna might increase environmental grain so that a higher meiofauna diversity could be maintained: for example, in the Wadden
Sea, Reise and Ax (1979) showed that certain meiofauna species have
definite preferences for well-defined sections of <u>Arenicola</u> burrows.

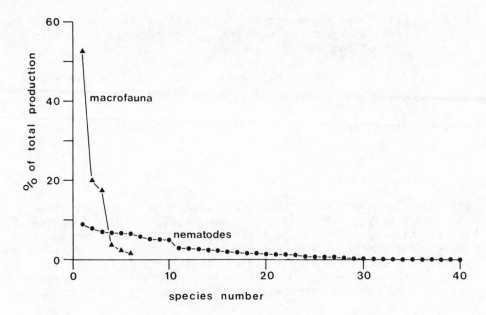

Figure 1. Partitioning of annual production amongst macrofauna species and meiofaunal nematode species from a mud-flat in the Lynher estuary.

However, no such structuring by permanent macrofauna burrows is evident in the Lynher.

The implications are, therefore, that in order for forty species of nematodes to co-exist in 1 ml of sediment (as they do in the Lynher), the sedimentary environment must be highly structured on a microscopic scale. If we observe fresh sediment through a stereo-microscope, for example through the walls of an aquarium, the nematodes (usually the most numerous and diverse taxon of the meiofauna) can be seen gliding along an intricate network of thread-like burrows which are reinforced by mucus secretions (Cullen, 1973). The mucus, secreted from the pharyngeal and caudal glands, may trap organic particles and adsorb macromolecules and thus act as source points for the growth of microorganisms, on which the nematodes feed (Riemann and Schrage, 1978). This "mucus-trap hypothesis", regarded as a "gardening" mechanism by Gerlach (1978) provides a method by which the nematodes could concentrate food but does not explain how the species partition resources. The fact that this partitioning appears to be so equitable poses a further problem, because it inevitably implies that some of the nematode species must be getting a good living out of relatively rare food species, and highly efficient foraging strategies would need to evolve in order that these food items could be exploited.

"Gardening" evolved in the neolithic period as a means of improving man's foraging strategy. Simply fertilising a piece of land will not dramatically improve the strategy because the preferred food will still be present in the same proportions as sub-optimal and inedible foods. The essence of efficient gardening is to provide a high density monospecific culture of the preferred food: in this way time spent in sorting and handling suboptimum foods is eliminated and the time spent between one preferred food item and the next is reduced to a minimum. If meiofauna were able to "garden" in this sense, it would enable them to exploit food items which might be rare when their average density (m^{-2}) is considered, but might be abundant in their gardens.

Some recent experiments in which we have attempted to culture nematode species on natural mixtures of micro-organisms from the mud-flat have suggested that highly efficient and specific mechanisms may exist. When the nematode Praeacanthonchus punctatus (Bastian) was introduced to such a micro-organism mix in sloppy nutrient agar, their trails were found after some while to comprise very high monospecific concentrations of spherical photosynthetic cells, on average 14.4 μm in diameter (Fig. 2). These were subsequently found to be the non-motile phase of the flagellate chlorophyte Tetraselmis, the motile phase of which was found randomly in the medium. Controlled experiments with monospecific cultures of the motile phase showed that the presence of Praeacanthonchus induced the formation of these non-motile cells, whereas in cultures without nematodes all the cells remained motile over the same time period (R. Warwick & P.C. Reid, unpublished). Nematodes of this type are known to feed by taking a cell into their buccal cavity, puncturing it with a cuticular tooth (or teeth), sucking out and ingesting the contents and rejecting the cell wall. The non-motile cells of Tetraselmis fit exactly into the buccal cavity of Praeacanthonchus (Fig. 3) and so, morphologically, would be the optimal food. We also know that such resting cells, in addition to being unable to escape from the nematode, may have a higher energy content than the motile cells (Droop, 1955) and may also be richer in lipids (Czygan, 1968; Sprey, 1970). Whether the nematode trails simply act as a "fly-paper" for concentrating Tetraselmis, or whether they attract and encourage active growth and multiplication of the alga is not known, but whatever the mechanism, the final concentrations of Tetraselmis in the tracks are very high and no other species are present.

It is possible that such mechanisms could be extensive in marine sediments but they are very difficult to investigate analytically, since any method of analysis inevitably destroys the physical, chemical and biological structure on which they must depend. The reason why most of the common nematode species cannot be maintained in laboratory culture may well be that it is impossible to replicate the complex environmental conditions under which these mechanisms

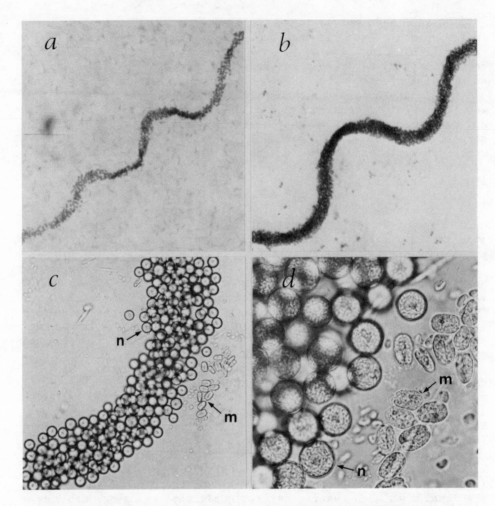

Figure 2. Trails of the nematode <u>Praeacanthonchus</u> <u>punctatus</u> in
sloppy nutrient agar with a natural mixture of micro-organisms from
the Lynher mud-flat. a, viewed through the stereomicroscope showing
early development of concentrations of non-motile <u>Tetraselmis</u> cells.
b, a later stage of concentration. c, viewed through the compound
microscope showing motile (m) and non-motile (n) phases of <u>Tetra-</u>
<u>selmis</u>. d, a detail from Fig. c at higher magnification. Non-motile
<u>Tetraselmis</u> cells are 13.5 - 16.5 μm in diameter.

can operate. Reports of these phenomena are therefore scant, but
Lee (1974), for example, described how another chlorophyte
<u>Dunaliella</u> stimulated rapid pseudopodial formation in three fora-
miniferan species, and how the flagellates ceased random motion and
swam directly to the pseudopods where they stopped swimming and were

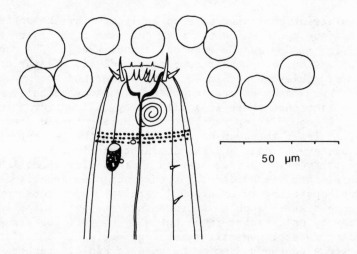

Figure 3. Head-end of <u>Praeacanthonchus</u> <u>punctatus</u> drawn to same
scale as non-motile <u>Tetraselmis</u> cells. Note exact correspondence
in size between <u>Tetraselmis</u> cells and the nematode buccal cavity.

lysed. Also, nematodes of the sub-family Stilbonematinae are known
to carry monospecific "gardens" of blue-green bacteria along as a
dense cuticular covering (see Gerlach, 1978).

One can envisage at least some conservative elements of the
meiofauna having these intimate and specific relationships with the
microbial populations, and maintaining high diversity and stability.
It is clear that this high level of organisation is quick to recover
from even severe physical disturbances (e.g. Sherman and Coull, 1980).
and the species involved do not seem to be exploited by macrofaunal
predators: at least we have been unable to identify significant
numbers of nematodes in the guts of several potential predators of
meiofauna in the Lynher.

This is not to say, however, that all elements of the meiofauna
are equally conservative and stable and maintain their diversity and
stability in the same way. The second most abundant and diverse
meiofauna group on the Lynher mudflat, the harpacticoid copepods,
form a significant item in the diet of several mobile predators,
particularly gobies. There are eight species on this mudflat, each
having marked seasonal fluctuations in abundance (Teare, 1978).
Obviously, there is a higher risk of predation in some types of
foraging than others (Maynard Smith, 1978), and we find that four
active epibenthic harpacticoid species comprise the preferred food
of gobies and that their seasonal abundance cycles are inversely
correlated with the abundance cycles of the gobies. On the other

hand, one sluggish burrowing copepod is ignored by the gobies and
has an abundance cycle which peaks at the same time as the gobies
(Teare, 1978; M. Gee and R. Warwick, unpublished). The preferred
copepod species build up their numbers rapidly in the spring and
early summer: gobies appear in the late summer and autumn when the
numbers of these copepods are immediately and dramatically reduced
despite the fact that there is still plenty of food for them (Joint,
1978). These copepods apparently have a more catholic diet than the
nematodes, at least they can be maintained in the laboratory on a
number of different foods and the dominant species, Tachidius
discipes Giesbrecht, thrives best on a mixed diet (Smol & Heip,
1974). Here, there is evidence for the mechanism described by
Paine (1966) operating to maintain diversity: predators apparently
regulate densities to such an extent as to prevent monopolisation
of resources by one species in potentially competitive situations,
thus enabling several generalist feeding species to co-exist. In
the Lynher, the copepods seem to be able to exploit rather rapidly,
short-term surpluses of a variety of foods which are not available
to, or cannot be exploited quickly enough by, the more conservative
nematodes, and in this sense appear to be more opportunistic. It
should not be thought, however, that conservativeness or opportunism
follow rigid taxonomic boundaries. There appears, for example, to
be a reserve of opportunistic nematodes present in the mud-flat,
although none of these are very abundant. They exist, perhaps as
resting eggs or very low densities of adults, ready to exploit new
situations should environmental conditions change. When artificial
agar-based media are placed on the mud-flat for 24 hours and then
brought back to the laboratory, cultures of nematode species (mainly
monhysterids and chromadorids) become established and these are
invariably either rare in the wild, or in some cases have never
been found in field samples at all. Perhaps they are generalist
feeders and their feeding conditions are more easily satisfied on
artificial media than those of the specialist "gardeners".

METABOLIC STRATEGIES

     The development of a conservative or opportunistic strategy
must be accompanied by biochemical, physiological and behavioural
adaptations which will allow these strategies to be adopted. In
order to maintain efficiency of resource exploitation in a seasonal
environment, species in these two categories should adapt their
metabolic functions differently in relation to temperature. Opportunists must be able to exploit a short-term food surplus fully, and
conversely reduce their requirements at times of food shortage. High
food levels are almost always associated with higher temperatures,
so that high Q_{10} values for metabolic functions might be expected
for opportunists and conversely, if a conservative species is in
balance with a relatively stable food resource it would be a disadvantage for it to increase its metabolic activity, since energy
demands might then exceed the supply. Our own data on respiration

rates of meiofauna species in relation to temperature (Price and Warwick, 1980), together with other data from the literature, suggest that there is a dichotomy in Q_{10} values: species exploiting season-ally unstable food resources have a Q_{10} of about two, over the range of temperatures they are likely to experience in the field, whereas species exploiting seasonally stable food resources have a Q_{10} of about one, i.e. their respiration rate is unaffected by temperature.

Such comparisons are facilitated by the fact that respiration measurements can be made on both conservative and opportunistic species taken from the field. However, comparisons of other temp-erature dependent functions are more difficult to study. For example, the intrinsic rate of natural increase (r), involving growth rate and fecundity, should similarly be temperature dependent in opportunists but not conservative species, but such comparisons are hampered by the fact that the conservative species cannot be cultured in the laboratory. All we can say at present is that the generation times of species in culture is markedly temperature dependent, (e.g. Hopper et al., 1973; Smol and Heip, 1974), and there are indications, too, that fecundity may increase with temp-erature. The example in Fig. 4 shows the life cycle of an opport-unistic nematode, Chromadora nudicapitata (Bastian), isolated from the Lynher mud-flat, in culture at 20°C and 15°C. The reduction in temperature by 5°C results in an increase in generation time (egg to egg) from 13 to 19 days. However, each developmental phase is not extended proportionally; the period from egg deposition to hatching is increased from 2 to 7 days whereas the period from hatching to egg production is only extended by one day from 11 to 12 days. This latter phase is perhaps the most susceptible to predation, so that the alteration of the relative lengths of develop-mental stages may have some adaptive significance. Perhaps more importantly, the fecundity is dramatically reduced: at 20°C the female lives for 6 days, during which time she produces more than 50 eggs, or 8.5 eggs per day, but at 15°C she lives for 9 days and produces only 8 eggs, or less than one per day.

REPRODUCTIVE AND LIFE-CYCLE STRATEGIES

Reproductive strategies in meiofauna are adopted, in combination with feeding and metabolic strategies, to optimise the number of offspring which can efficiently utilise available resources. Certain conservative features of the reproductive biology of meiofauna in-herently lead to community stability (Warwick, 1980). These include fertilisation by copulation, direct benthic development, brood protection, viviparity, parthenogenesis, hermaphroditism, continuous reproduction etc. exhibited with varying degrees of generality throughout the meiofaunal taxa. Dispersive capability is sacrificed, but the animals are sufficiently small to be dispersed as adults by water currents (Gerlach, 1977). Of particular importance are those adaptation which enable the meiofauna to maintain population levels

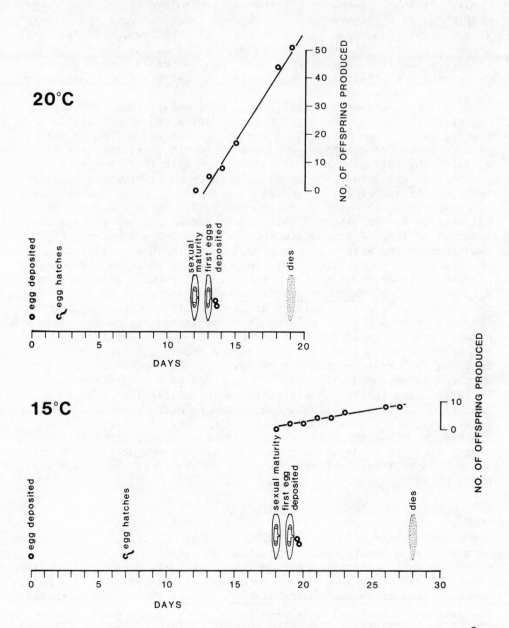

Figure 4. Life cycles of a female <u>Chromadora</u> <u>nudicapitata</u> at 20°C and 15°C.

appropriate to food levels in a seasonal environment. In contrast
to the macrofauna, meiofauna species have life cycles which are
shorter than the seasonal cycle so it is not necessary for those
species which depend on seasonally fluctuating resources to survive
as actively feeding adults during periods of shortage, when they
can remain physiologically quiescent. These periods might be spent
as resistent eggs (sometimes with a true diapause) or quiescent
(sometimes hibernating) adults, or in protective cysts (Warwick
1980).

A further consequence of the short generation time is that
certain density dependent mechanisms of population regulation can
operate effectively on the time scale of resource fluctuation. For
copulating species, providing the most efficient sex ratio is
important. In species which reproduce purely sexually, and where
the male is active in seeking out the more passive females, a ratio
in favour of females is advantageous at high densities, and in
favour of males at low densities. Sex ratio switching of this kind
in response to fluctuations in population density has been noted in
certain copepods both in the field (e.g. Penter, 1969) and under
laboratory conditions (e.g. Battaglia, 1964). On the other hand,
Clark (1978), making similar studies on the nematode Diplenteron
potohikus (Yeates), which showed facultative parthenogenesis, found
that at low densities all eggs gave rise to females i.e. males were
not produced when their probability of finding a mate was low, and
parthenogenetic reproduction was then employed. It has also been
found that, for certain copepods, fecundity is reduced with increased
crowding (Fava and Crotti, 1979; Walker, 1979). In both these
studies, and in that of Clark (1978), external metabolites or ecto-
crines have been implicated in the mediation of the response.

CONCLUSIONS

If a single conclusion emerges from the above discussion, it
must be that the organisation of meiofauna communities is so complex,
and attempts to analyse it so few, that we can only begin to specu-
late on overall survival strategies. What is clear, however, is
that such strategies cannot be considered by studying the meiofauna
in isolation, but that due consideration must be paid to their
interactions with trophic levels both above and below them. It is
equally clear that the categories of adaptation I have discussed
above cannot be considered in isolation: feeding, metabolic and
reproductive strategies are intimately dependent on one another to
provide an overall strategy for survival.

ACKNOWLEDGEMENTS

This work forms part of the estuarine ecology programme of the
Institute for Marine Environmental Research, a component institute
of the Natural Environment Research Council.

REFERENCES

Anderson, J.G. and Meadows, P.S. 1978. Microenvironments in marine
    sediments. Proc. Roy. Soc. Edin. 76B: 1-16.
Battaglia, B. 1964. Advances and problems of ecological genetics
    in marine animals. p. 451-461. In: Genetics Today. Pergamon
    Press.
Calow, P. 1977. Ecology, evolution and energetics: a study in
    metabolic adaptation. Adv. ecol. Res. 10: 1-62.
Caswell, H. 1976. Community structure: a neutral model analysis.
    Ecol. Monogr. 46: 327-354.
Clark, W.C. 1978. Metabolite - mediated density dependent sex
    determination in a free-living nematode, Diplenteron potohikus.
    J. Zool. Lond. 184: 245-254.
Coull, B.C. and Vernberg, W.B. 1975. Reproductive periodicity of
    meiobenthic copepods: seasonal or continuous? Mar. Biol.
    32: 289-293.
Cullen, D.J. 1973. Bioturbation of superficial marine sediments by
    interstitial meiobenthos. Nature, 242: 323-324.
Czygan, F-C. 1968. Sekundär-Carotinoide in Grünalgen. I. Chemie,
    Vorkommen und Faktoren, welche die Bildung dieser Polyene
    beinflussen. Arch. Microbiol. 61: 81-102.
Droop, M.R. 1955. Some factors governing encystment in Haematococcus
    pluvialis. Arch. Microbiol. 21: 267-272.
Fava, G. and Crotti, E. 1979. Effect of crowding on nauplii produc-
    tion during mating time in Tisbe clodiensis and T.holothuriae
    (Copepoda, Harpacticoida). Helgoländer wiss Meersunters. 32:
    466-475.
Gerlach, S.A. 1977. Means of meiofauna dispersal. Mikrofauna
    Meeresboden, 61: 89-103.
Gerlach, S.A. 1978. Food-chain relationships in subtidal silty sand
    marine sediments and the role of meiofauna in stimulating
    bacterial productivity. Oecologia, 33: 55-69.
Gray, J.S. 1978. The structure of meiofauna communities. Sarsia,
    64: 265-272.
Hopper, B.E., Fell, J.W. and Cefalu, R.C. 1973. Effect of temperature
    on life cycles of nematodes associated with the mangrove
    (Rhizophora mangle) detrital system. Mar. Biol. 23: 293-296.
Hutchinson, G.E. 1959. Homage to Santa Rosalia or why are there so
    many kinds of animals? Am. Nat. 93: 145-159.
Hutchinson, G.E. and MacArthur, R. 1959. A theorectical ecological
    model of size distribution among species of animals. Am. Nat.
    93: 117-126.
Joint, I.R. 1978. Microbial production of an estuarine mudflat.
    Est. coastl. mar. Sci. 7: 185-195.
Lee, J.J. 1974. Towards understanding the niche of foraminifera.
    p. 207-260. In: R.H. Hedley & C.G. Adams (eds.), Foraminifera
    Volume 1. Academic Press.
Lee, J.J., Tietjen, J.H., Mastropaolo, C. and Rubin, H. 1977.
    Food quality and heterogeneous spatial distribution of meio-

fauna. Helgoländer wiss. Meeresunters. 30: 272-282.

Levins, R. 1968. Evolution in changing environments. Some theoretical explorations. Monographs in Population Biology, 2: Princeton University Press.

MacArthur, R.H. 1955. Fluctuations of animal populations and a measure of community stability. Ecology 36: 533-536.

MacArthur, R.H. and Levins, R. 1964. Competition, habitat selection and character displacement in a patchy environment. Proc. nat. Acad. Sci. 51: 1207-1210.

McIntyre, A.D. 1971. Control factors on meiofauna populations. Thalassia Jugoslavica, 7: 209-215.

Maynard Smith, J. 1978. Optimization theory in evolution. Ann. Rev. Ecol. Syst. 9: 31-56.

Paine, R.T. 1966. Food web complexity and species diversity. Am. Nat. 100: 65-75.

Penter, D.M. 1969. The ecology of intertidal harpacticoids in the Swale, North Kent. Ph.D. Thesis, University of London.

Price, R. and Warwick, R.M. 1980. The effect of temperature on the respiration rate of meiofauna. Oecologia, 44: 145-148.

Reise, K. and Ax, P. 1979. A meiofaunal "thiobios" limited to the anaerobic sulfide system of marine sand does not exist. Mar. Biol. 54: 225-237.

Riemann, F. and Schrage, M. 1978. The mucus-trap hypothesis on feeding of aquatic nematodes and implications for biodegradation and sediment texture. Oecologia 34: 75-88.

Sherman, K.M. and Coull, B. 1980. The response of meiofauna to sediment disturbance. J. exp. mar. Biol. Ecol. 46: 59-71.

Sanders, H.L. 1968. Marine benthic diversity: a comparitive study. Am. Nat. 102: 243-282.

Simberloff, D. 1972. Properties of the rarefaction diversity measurement. Am. Nat. 106: 414-418.

Smol, N. and Heip, C. 1974. The culturing of some harpacticoid copepods from brackish waters. Biol. Jb. Dodonaea, 42: 159-169.

Sprey, B. 1970. Die Lokalisierung von Sekundärcorotinoiden von Haematococcus pluvialis Flotow em. Wille. Protoplasma, 71: 235-250.

Teare, M.J. 1978. An energy budget for Tachidius discipes (Copepoda Harpacticoida) from an estuarine mud-flat. Ph.D. Thesis, University of Exeter.

Tietjen, J.H. and Lee, J.J. 1977. Feeding behaviour of marine nematodes. p. 21-36. In: B.C. Coull (ed.), Ecology of marine benthos. University of South Carolina Press.

Walker, I. 1979. Mechanisms of density-dependent population regulation in the marine copepod Amphiascoides sp. (Harpacticoida). Mar. Ecol. Prog. Ser. 1: 209-221.

Warwick, R.M. 1980. Population dynamics and secondary production of benthos. p. 1-24. In: K.R. Tenore and B.C. Coull (eds.), Marine benthic dynamics. University of South Carolina Press.

Warwick, R.M., Joint, I.R. and Radford, P.J. 1979. Secondary

production of the benthos in an estuarine environment. p. 429-
450. In: R.L. Jefferies and A.J. Davey (eds.), Ecological
Processes in Coastal Environments. Blackwell.
Warwick, R.M. and Price, R. 1975. Macrofauna production in an
estuarine mud-flat. J. mar. biol. Ass. U.K. 55: 1-18.
Warwick, R.M. and Price, R. 1979. Ecological and metabolic studies
on free-living nematodes from an estuarine mud-flat. Est.
coastl. mar. Sci. 9: 257-271.

SURVIVAL STRATEGIES OF TUBIFICIDS IN THE THAMES AND OTHER ESTUARIES

Julian Hunter

Highland River Purification Board, Strathpeffer Road
Dingwall, Scotland
formerly of: Zoology Department, King's College
University of London

INTRODUCTION

Eight tubificid species form a succession along the Thames estuary which is similar to that found in some other British & European estuaries (Fig. 1). The freshwater species comprise

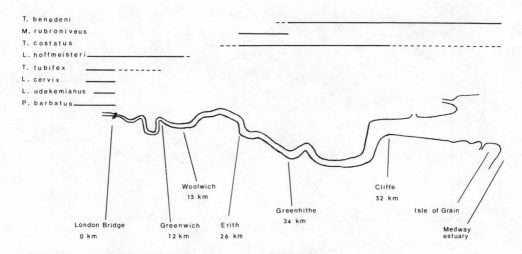

Figure 1. Thames estuary. Distances are seaward of London Bridge. A solid line denotes the continuous presence of the species; a hatched line denotes seasonal decline or elimination. The exact limits of each species are changeable and approximate.

<u>Limnodrilus</u> <u>hoffmeisteri</u> Claparède, <u>Tubifex</u> <u>tubifex</u> Müller,
<u>Limnodrilus</u> <u>udekemianus</u> Claparède, <u>Limnodrilus</u> <u>cervix</u> Brinkhurst
and <u>Psammoryctides</u> <u>barbatus</u> Grube. This freshwater component of
the fauna extends seawards to London Bridge where the salinity of
the water reaches a maximum of about 5‰.

    At Greenwich, 12 km seaward of London Bridge, <u>L.hoffmeisteri</u>
was continuously present between 1967 and 1973 while <u>T.tubifex</u>
appeared seasonally. <u>L.hoffmeisteri</u> was collected at Woolwich
whenever samples were taken, although no regular programme was undertaken here.

    Figure 2 shows typical salinity and dissolved oxygen profiles
for a period of fairly low freshwater in September 1972. At Erith
salinity fluctuated between 2‰ and 20‰ and this area suffered very
low dissolved oxygen conditions due to the proximity of the main
sewage outfalls. This salinity range was unsuitable for any but the
most euryhaline species and the fauna was dominated by <u>Tubifex</u>
<u>costatus</u> Claparède with much lower numbers of <u>Monopylephorus</u>
<u>rubroniveus</u> Levinsen.

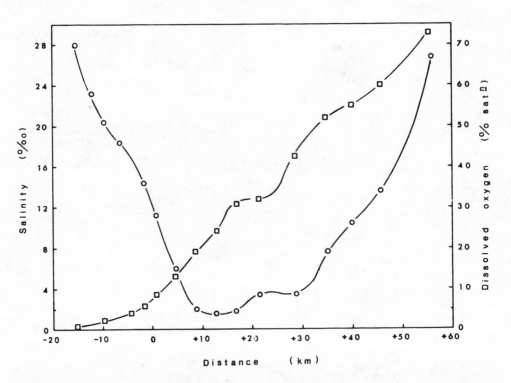

Figure 2. Typical salinity and dissolved-oxygen profiles for the
Thames estuary in September 1972. Distances seaward of London
Bridge: o——o , dissolved-oxygen: □——□ , salinity.

At Greenhithe, T.costatus was still the dominant species but
the marine tubificid Tubificoides benedeni Udekem was a minor
component of the fauna. At Cliffe, T.benedeni had become the dominant species and T.costatus had become less significant. The cirra-
tulid polychaete Caulleriella zetlandica McIntosh was also recorded
at Cliffe.

THE FRESHWATER SPECIES

Some reduction in tubificid diversity occurred between Reading
in the freshwater zone where Kennedy (1964a) recorded fifteen species,
and London Bridge, where six species were recorded (Hunter, 1977),
but the reason for this is not known. Of the five species regularly
collected at London Bridge only L.hoffmeisteri and T.tubifex were
found at Greenwich in any numbers. L.cervix is an exotic species
spreading through canals and associated rivers (Kennedy, 1964a).
It has not been included in species lists from British estuaries so
its elimination seawards of London Bridge is probably due to salinity.

P.barbatus has been recorded in salinities at least as high as
T.tubifex (Timm, 1970) so its limit at London Bridge may be due to
other factors. Brinkhurst (1965b) found it tolerant of pollution
as long as the water was well aerated, so perhaps decreasing dissolved oxygen seawards of London Bridge may limit this species.
L.udekemianus is less tolerant of salinity than L.hoffmeisteri
(Kennedy, 1964b) and the limited information in Table 1 suggests
that salinity may be restricting its distribution.

T.tubifex and L.hoffmeisteri are two of the commonest freshwater species and a considerable amount of work has been done on
their distribution and ecophysiology. Birtwell (1972) tested the
salinity tolerance of adult T.tubifex at 20°C and found a 168h LC_{50}
of 9.0‰ while Hunter (1977) tested the survival of newly hatched
worms and found a 168h LC_{50} of 6.7‰. This species would be expected
to survive the range of salinity at London Bridge but not the extremes
of salinity at Greenwich in the summer and autumn. Field results
(Hunter, 1977) show that T.tubifex was eliminated from the Greenwich
fauna in autumn 1970 and 1972 when salinity was between 6‰ and 10‰.
In autumn 1971, when salinity was generally 4‰-6‰, this species
declined in numbers but persisted through to the winter. The effects
of salinity on the reproductive cycle of T.tubifex were also tested
by exposing breeding worms in natural sediment to a range of salinities for 14 days (Hunter, 1977). Production of cocoons and embryos
was significantly reduced at 5‰ compared with the control at 0.7‰.
At London Bridge, breeding in T.tubifex was restricted during 1970
and 1971 to the period January to May, when estuary salinity was
low. This contrasts with the life cycle in freshwater, as Brinkhurst
(1965a) considered that T.tubifex was able to colonise polluted water
due to its ability to reproduce throughout the year. Ladle (1971)
also reported an extended reproductive period in two Dorset streams.

TABLE 1   Limiting Salinities for some Freshwater Tubificids
          Penetrating Estuaries

LOCATION	LIMITING SALINITY	AUTHORITY
	Limnodrilus hoffmeisteri	
Thames estuary	10‰ -12‰	Hunter, 1977
Tees estuary	9‰	Alexander et al, 1935
Baltic Sea	3‰ -4‰	Leppakoski, 1967
San Francisco Bay	mean 7‰ , max 12-13‰	Brinkhurst & Simmons, 1968
Baltic Sea	6‰	Laakso, 1969
Baltic Sea, Finland	4.3‰	Bagge and Ilus, 1973
Schlei fjord, Baltic Sea	8‰	Pfannkuche, 1974
Forth estuary	4‰	McLusky et al, 1980
	Tubifex *tubifex*	
Tees estuary	3‰	Alexander et al, 1935
Thames estuary	5‰ -7‰	Hunter, 1977
Forth estuary	4‰	McLusky et al, 1980
Baltic Sea	3‰	Laakso, 1969
Baltic Sea	3‰	Timm, 1970
	Psammoryctides *barbatus*	
Baltic Sea	5‰	Bagge and Ilus, 1973
Baltic Sea	6‰ -7‰	Laakso, 1969
Baltic Sea	7‰	Timm, 1970
Schlei fjord, Baltic Sea	8‰	Pfannkuche, 1974
	Limnodrilus udekemianus	
Thames estuary	5‰	Hunter, 1977
Baltic Sea	3‰	Bagge and Ilus, 1973
Baltic Sea	6‰	Laakso, 1969
Schlei fjord, Baltic Sea	8‰	Pfannkuche, 1974

Occasional collections of T.tubifex taken by myself in the R. Lee, Hackney, showed in a high proportion breeding in autumn 1970 when no breeding worms were found at London Bridge.

L.hoffmeisteri was reported by Kennedy (1964b) to be more tolerant of salinity than other species of the genus. In laboratory experiments, Birtwell & Arthur (1980) reported a 168h LC_{50} for adults of 14.7‰ while the corresponding figure for newly-hatched worms was 11.8‰ (Hunter, 1977).

The population studies cited in Table 1 suggest a limiting salinity of about 9‰ -10‰ and although this figure was exceeded for short periods in the autumn months this species was continuously present at Greenwich from 1967 to 1973. No regular sampling was undertaken at Woolwich, but as estuary salinity could reach 12‰ some deleterious effects on the L.hoffmeisteri population would be expected. In experiments on the reproductive cycle, cocoon and embryo production was inhibited at salinities of 5‰ and above. The life cycle of L.hoffmeisteri was followed over a three year period at Greenwich, (Hunter, 1977) and in each year the majority of worms reached the breeding stage in winter and early spring, with the peak of cocoon density in January and February. In contrast, this species showed "fairly uniform reproductive activity throughout the year" in the freshwater R. Thames (Kennedy, 1966). Thus, the breeding season in both L.hoffmeisteri and T.tubifex appears to be restricted under estuarine conditions compared with freshwater conditions. This supports the view of Kinne (1966) that "in freshwater species pene- trating estuaries, salinity may modify the time and length of the breeding season".

Plankton hauls made in the Thames estuary showed considerable numbers of freshwater tubificids of several species being carried seawards in the water column, presumably eroded from the sediment by surf and tidal currents. Their eventual distribution would depend on the salinity and other conditions encountered when they resettled, so that populations would extend further seaward in conditions of high freshwater flow and low salinity. Conversely the marine and estuarine species extended further landward in low-flow conditions, probably with the assistance of saline bottom currents, but resumed a more seaward distribution in winter. Brinkhurst and others (pers. comm.) have also found that the fauna of the Fraser estuary, Canada, moves seawards and landwards in a seasonal manner. Benthic tubificids may thus have a more dynamic distribution in estuaries than is often realised.

THE BRACKISH-WATER SPECIES

The tolerance of T.costatus to reduced salinities was tested by Birtwell and Arthur (1980) who reported an LC_{50} at 20°C of 0.5‰ . This fits in well with its limit just landward of Erith where estuary

TABLE 2 Limiting Salinities for some Estuarine and Marine Tubificids

LOCATION	LIMITING SALINITY	AUTHORITY
	Tubifex costatus	
Thames estuary	>1‰ [a], <21‰ [b]	[a]Birtwell and Arthur, 1980 [b]Hunter and Arthur, 1978
Schlei fjord, Baltic Sea	>1‰	Pfannkuche, 1974
Baltic Sea, Finland	>1‰	Laakso, 1969
Forth estuary	>2‰, <26‰	McLusky et al, 1980
	Monopylephorus rubroniveus	
Thames estuary	>2‰, <25‰ approx.	Hunter, unpublished data
Schlei fjord, Baltic Sea	>4‰, <12‰ approx.	Pfannkuche, 1974
	Tubificoides benedeni	
Thames estuary	>4‰ -7‰	Hunter, 1977
Schlei fjord, Baltic Sea	>7‰ approx.	Pfannkuche, 1974
Tees estuary	>6‰	Alexander et al, 1935
Danish estuaries	>12‰	Muus, 1967
Forth estuary	>2‰	McLusky et al, 1980

salinity fell to about 1‰ in winter, (Table 2). These workers also
found a fairly prolonged breeding season between January and June at
Erith and over this period estuary salinity increased from 1‰ to
about 9‰ . They also found that T.costatus tolerated a salinity of
34‰ with no mortality, so it is surprising that this species declined
seaward of Greenhithe. At Cliffe, the density of T.costatus was highest when the estuary salinity was lowest and, in general 21‰ was
limiting for significant populations of this species (Hunter and
Arthur, 1978). Similarly, in the Medway estuary Wharfe (1977) did
not record T.costatus from the more marine area around the Isle of
Grain. Bagge (1969) found T.costatus to be dominant in the mesohaline
reaches of the Saltkallfjord while T.benedeni became dominant further
seaward. In the Forth estuary (McLusky et al., 1980), the main
T.costatus population was upstream of Kincardine Bridge where interstitial salinity was between 3‰ and 26‰ , which is very similar to

its salinity range in the Thames estuary. Seaward of Kincardine
Bridge, T.benedeni replaced T.costatus as the dominant species as it
did in the Thames estuary.

Although able to survive in full seawater, T.costatus generally
appears to dominate the fauna in areas of reduced or fluctuating
salinity and gives way to other species, particularly T.benedeni
and polychaetes, at salinities over about 26‰. As Jansson (1962)
showed, the range of salinity preference of oligochaetes may be much
narrower than their range of tolerance.

M.rubroniveus also occurred in low numbers at Erith and Green-
hithe, and as it has been recorded in other estuaries, e.g. the
Clyde (Henderson, unpublished data), a brief mention is made in
Table 2.

THE MARINE SPECIES

This component of the fauna is represented in the Thames by
T.benedeni, a species often recorded from coastal and offshore
waters as well as estuaries. Populations penetrated the estuary to
Greenhithe where they constituted 5-10% of the fauna and here winter
salinity was typically 7-9‰ with a minimum of about 4‰. In salinity
tolerance experiments, Birtwell and Arthur (1980) reported an LC_{50}
of 5.9‰ at 20°C and 2.8‰ at 5°C which fits in well with its distribution in the estuary. They also showed that T.benedeni has an
exceptional ability to tolerate anoxia without incurring an oxygen
debt, so that it could perhaps retreat into the sediment to avoid
unsuitable salinity conditions in the estuary.

The life-cycle of T.benedeni is described by Hunter and Arthur
(1978). The main breeding period at Cliffe in 1972 was from April
to August and in 1973 it was from April to July. During this period,
the salinity at Cliffe was generally in the upper part of the range.
In the Forth estuary, McLusky et al. (1980) considered this species
to be the most abundant and the most important in terms of biomass.
The same would probably apply to the Thames estuary. Several other
marine and brackish-water tubificids which have been recorded in
other estuaries were not found in the Thames. For instance,
Tubificoides pseudogaster Dahl is common in the Forth estuary
(McLusky et al., 1980) and in the Clyde estuary, (Henderson, unpublished data). Aktedrilus monospermathecus Knöllner was found in a wide
range of salinities in the Schlei fjord (Pfannkuche, 1974) and is the
dominant littoral tubificid in the Cromarty Firth, (Hunter, unpublished data).

TUBIFICIDS AND DISSOLVED-OXYGEN

Little mention has been made of dissolved-oxygen in this paper
despite the polluted nature of the Thames estuary. This is because

Hunter (1977) and Birtwell and Arthur (1980) considered salinity to
be the "Master factor" (Kinne, 1966) affecting tubificid distribution
with dissolved-oxygen acting as an "Accessory factor". In the better
aerated reaches of the estuary at, and above, London Bridge the greatest density of worms was at low tide level whereas in the more deoxygenated reaches at Woolwich, Erith and Greenhithe the density of
worms increased up the beach. It is possible that worms could
extract oxygen from the surface water film while the shore was exposed. Fisher and Beeton (1975) found that L.hoffmeisteri would
move from sediment covered by deoxygenated water to sediment covered
by aerated water.

During the autumn of 1970 there was an extended period when
dissolved-oxygen was very low (about 1% saturation) and this was
thought to be responsible for retarded growth and maturation in
T.costatus by Birtwell and Arthur (1980) and in L.hoffmeisteri by
Hunter (1977). In laboratory experiments (Hunter, 1977), production
of cocoons and embryos of L.hoffmeisteri and T.tubifex was inhibited
by concentrations of dissolved-oxygen well above lethal levels so
that this factor may play a part, with salinity, in restricting
reproduction to the winter and early spring months. Anoxic conditions
have not occurred in the Thames estuary for a number of years but
in the upper Clyde estuary the water frequently becomes anoxic in
summer, eliminating the benthic tubificids which then recolonise the
estuary in winter (Mackay et al., 1978).

VERTICAL DISTRIBUTION IN THE SEDIMENT

The vertical distribution of T.benedeni was studied at Cliffe
(Hunter and Arthur, 1978). In the population as a whole, the greatest proportion of worms was in the superficial 2 cm but the greatest
proportion of both breeding worms and cocoons was in the 2-4 cm layer,
which was often black and anoxic (Fig. 3). In this situation the
cocoons may be more protected against erosion and extremes of salinity
and temperature. However, the newly-hatched worms, which in experiments were more sensitive to reduced dissolved-oxygen, were found in
the superficial layer. A similar pattern was found at Greenwich
where cocoons of L.hoffmeisteri occupied the deeper layers.

Seasonal trends were also seen in the vertical distribution of
tubificids. Birtwell and Arthur (1980) reported a movement towards
the sediment surface by T.costatus during the warmer months.
L.hoffmeisteri was also found closer to the surface during the summer
(Hunter, 1977), and these two species probably move upward to avoid
the black reduced sediments which extend closer to the surface in
summer. In T.benedeni, similar clear seasonal migrations were not
found (Hunter and Arthur, 1978), possibly because of the great tolerance of this species to anoxia.

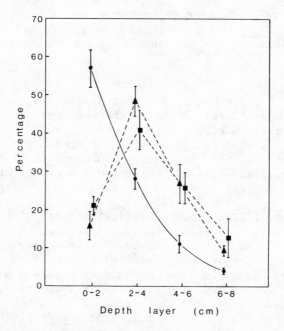

Figure 3. The vertical distribution of T.benedeni during the
reproductive period in 1973: mean percentage and standard error.
■ , cocoons; ▲ , breeding worms; ● total worms.

SUMMARY

     Evidence from experiments and field studies suggests that
salinity is the "Master factor" dictating the distribution of species
along the Thames estuary and that this distribution is very dynamic.
Worms are continuously being carried into the estuary from freshwater and from the sea and being distributed according to their
tolerance of salinity. The main breeding periods of the various
species also appear to fit in with their salinity tolerance. In the
freshwater species, L.hoffmeisteri, the main breeding period was
December to March; in the euryhaline T.costatus, the breeding peak
was between January and June, while in the marine T.benedeni, breeding peaks were in May to August 1972 and March to July 1973. It was
not easy to evaluate the effect of the dissolved-oxygen sag in the
Thames estuary but in estuaries where anoxic conditions occur, such
as the Clyde estuary, the fauna may be eliminated seasonally. The
various species were distributed in the sediment according to their
tolerance of deoxygenation. The young immature worms occupied the
surface layer while the breeding adult worms, which tolerated lower
dissolved-oxygen concentrations, were found in deeper layers where
they laid cocoons. L.hoffmeisteri and T.costatus showed a seasonal
pattern in their vertical distribution which parallelled movements

of the black reduced layer so that most worms remained above the
black layer. T.benedeni, which in experiments had the greatest
tolerance of anoxia, did not show this seasonal migration pattern.

REFERENCES

Alexander, W.B., Southgate, B.A. and Bassindale, R. 1935. A survey
    of the River Tees. Part 2 - the estuary; chemical and
    biological. W.P.R.L. Tech. Paper No. 5.
Bagge, P. 1969. The succesion of the bottom fauna communities in
    polluted estuarine habitats. Limnologica, 7: 87-94.
Bagge, P. and Ilus, E. 1973. Distribution of benthic tubificids in
    Finnish coastal waters in relation to hydrography and pollution.
    Oikos. Suppl. 15: 214-225.
Birtwell, I.K. 1972. Ecophysiological aspects of tubificids in the
    Thames estuary. Ph.D. Thesis, University of London.
Birtwell, I.K. and Arthur, D.R. 1980. The ecology of tubificids in
    the Thames estuary with particular references to Tubifex
    costatus Claparède. In "Aquatic oligochaete biology", edited
    by Brinkhurst, R.O. & D.G. Cook. Plenum Press, 1980.
Brinkhurst, R.O. 1965a. Observations on the recovery of a British
    river from gross organic pollution. Hydrobiologia, 24: 9-51.
Brinkhurst, R.O. 1965b. The biology of the Tubificidae with special
    reference to pollution. In "Biological problems in water
    pollution", U.S. Dept. of Health, Publication No. 999 - WP - 25.
Brinkhurst, R.O. and Simmons, M.L. 1968. The aquatic oligochaete
    of the San Francisco Bay system. Calif. Fish & Game, 54:
    180-194.
Fisher, J.A. and Beeton, A.M. 1975. The effect of dissolved-oxygen
    on the burrowing behaviour of Limnodrilus hoffmeisteri
    (Oligochaeta) Hydrobiologia, 47: 273-290.
Hunter, J.B. 1977. Some aspects of the ecology of tubificids in the
    Thames estuary. Ph.D. Thesis. University of London.
Hunter, J.B. and Arthur, D.R. 1978. Some aspects of the ecology of
    Peloscolex benedeni Udekem (Oligochaeta: Tubificidae) in the
    Thames estuary. Estuar. cstl. mar. Sci., 6: 197-208.
Jansson, B.O. 1962. Salinity resistance and salinity preference of
    two oligochaetes from the interstitial fauna of marine sandy
    beaches. Oikos, 13: 293-305.
Kennedy, C.R. 1964a. The biology of some freshwater oligochaetes
    Ph.D. Thesis, University of Liverpool.
Kennedy, C.R. 1964b. Studies on the Irish Tubificidae. Proc. R.
    Irish Acad., 63B: 225.
Kennedy, C.R. 1966. The life history of Limnodrilus hoffmeisteri
    Clap. (Oligochaeta: Tubificidae) and its adaptive significance.
    Oikos, 17: 159-168.
Kinne, O. 1966. Physiological aspects of life in estuaries with
    special reference to salinity. Netherl. J. Sea. Res., 3:
    222-244.
Laakso, M. 1969. Oligochaeta from brackish water near Tvarminne,

S.W. Finland. Ann. Zool. Fenn., 6: 98–111.

Ladle, M. 1971. The biology of oligochaeta from Dorset chalk streams. Freshwat. Biol. 1: 83–97.

Leppakoski, E. 1967. Notes on the aquatic oligochaeta (Annelida) of the Bothnian Bay, Finland. Aquilo, (ser. zool.) 5: 30–34.

Mackay, D.W., Tayler, W.K. and Henderson, A.R. 1978. The recovery of the polluted Clyde estuary. Proc. Roy. Soc. Edin., 76B: 135–152.

McLusky, D.S., Teare, M. and Phizacklea, P. 1980. Effects of domestic and industrial pollution on the distribution and abundance of aquatic oligochaetes in the Forth estuary. Helgolander wiss Meeresunters. In Press.

Muus, B.J. 1967. The fauna of Danish estuaries and lagoons. Meddr. Danm. Fisk. og Havanders, 5: 7–316.

Pfannkuche, O. 1974. Zur systemtik und Okologie naidomorpher Brackwasseroligochaeten. Mitt Hamburg. Zool. Mus. Inst., 71: 115–134.

Timm, T. 1970. On the fauna of the Estonian oligochaeta. Pedobiologia, 10: 52–78.

Wharfe, J.R. 1977. An ecological survey of the benthic invertebrate macrofauna of the lower Medway estuary, Kent. J. Anim. Ecol., 46: 93–113.

SURVIVAL STRATEGIES IN ESTUARINE NEREIDS

C. Mettam

Department of Zoology
University College, P.O. Box 78
Cardiff, Wales

INTRODUCTION

Nereid polychaetes are among the most abundant worms in the estuaries of north-western Europe. Like most other estuarine macrofauna, they have their origins in marine conditions and the colonisation of estuaries has presented special challenges, both physiological and ecological. This paper considers some of the special strategies which appear to have accompanied the penetration and colonisation of estuaries by nereids.

Strategies are tactics employed to attain a particular, defined goal which, for living, reproducing organisms, is the perpetuation of their genes in future generations. Tactics which maximise surviving offspring are self perpetuating through genetic inheritance. This is the nature of Natural Selection. It follows that strategies are not purposeful behaviours on the part of animals: they can't help doing their best, as selection pressures demand, operating within environmental or inherent constraints.

Estuaries are a hostile environment in the sense that marine animals must pay some energetic cost to live there, although benefits may accrue from an abundance of food and lack of competition from other species. The rewards go to the species that can effect the best compromise between costs and benefits, or 'maximise the trade-off' in the jargon of current literature (Pianka, 1978). Four nereid species, Nereis (Hediste) diversicolor, N.(Neanthes) virens, N.(Neanthes) succinea, N.(Nereis) pelagica, commonly penetrate into European estuaries and show differences in their physiological tolerances (Theede et al., 1973) and in the pattern of their life cycles. Their ecology is reviewed by Wolff (1973).

65

REPRODUCTION

Nereis pelagica exemplifies the primitive, marine, dioecous
condition in that, after some time feeding and growing on the sea
bed, both sexes metamorphose into swarming epitokes, swim out to sea,
spawn and subsequently die. Some populations, however, may reproduce
as atokes, without metamorphosis (Rasmussen, 1956). Nereis succinea
is more euryhaline but retains the classic reproductive behaviour.
Nereis virens is characteristic of slightly reduced salinities and,
in some populations, possibly all, swarming is confined to the males,
females remain in their burrows (Bass & Brafield, 1972). Nereis
diversicolor remains atokous throughout its life, the changes corresponding to the morphological elaborations of heteronereids are confined to a modest histolysis of somatic tissues (Dales, 1950). It is
morphologically indistinguishable from two other non-european species,
N.limnicola which is a viviparous, self-fertilising hermaphrodite
penetrating into fresh water and the estuarine N.japonica which
swarms, without heteronereid metamorphosis, moving seawards to spawn.

The reproductive biology of N.diversicolor in the Thames estuary
at Chalkwell has been described in detail by Dales (1950). A surprising finding, in terms of evolutionary strategy, was the scarcity
of males in the breeding population, a widely recorded phenomenon
reaching an extreme at Chalkwell where less than 10% of sexually
mature worms were male. Sex ratios for animals with a simple chromo-
somal mechanism of sex determination are normally 1:1. The sex ratio
of N.japonica from field collections and artificial fertilisation
lies close to 1:1 (Inamori & Kurihara, 1979d).

There is also a strategic reason why sex ratios tend to equality.
In a panmictic, dioecious population, each sex has the same opportunity to transmit genes to the next generation; there is no advantage
to the individual investing more in offspring of either sex and the
sex ratio stabilises at 0.5 (Fisher, 1930). A departure from this
equilibrium may occur if sexual differentiation affects adult survival
(e.g. Nereis fucata matures in two and three years for males and
females respectively, leading to an excess of adult males (Goerke,
1971b)), but there are no obvious differences between the sexes of
N.diversicolor, other than in the gametes themselves.

Nereis diversicolor probably spawns in gregarious clusters on
the shore (Dales, 1950). In this intimate proximity, fertilization
must be more effective than in the open sea and fewer males are
needed. While the group might benefit by reducing the sex ratio,
there is still no advantage to the individual parent in producing an
imbalance of offspring and an innate tendency to produce females
will not persist because the production of males is then favoured,
restoring the balance.

Fisher's (1930) original argument concerned a balance of parental

investment in each sex. A period of parental care, an investment
that could otherwise have gone into gamete production, is made possible by the supression of pelagic reproduction. Brooding occurs in
several nereids (Mazurkiewicz, 1975) and possibly in N.diversicolor
but it seems unlikely that the imbalanced sex ratio arises at this
time. A secondary sex ratio of 0.2 requires the cost per male to be
4 x the cost per female and would require heavy losses of zygotes
during embryogenesis.

Nereis limnicola, as a viviparous, self-fertilising hermaphrodite
can protect gametes and early embryos from osmotic stress at low
salinities but has reduced opportunities for genetic recombination.
A tendency towards both parthenogenesis and viviparity has been found
in N.diversicolor, although not shown to be an effective means of
reproducing (Bogucki, 1953; Smith, 1976). As a strategy, parthenogenesis maximises inheritance of genes and the proportion of
parthenogenetic females will double with every bout of reproduction
(Maynard Smith, 1978). Dales (1950) reported that female
N.diversicolor would not spawn without a male present and that un-
spawned eggs degenerated eventually forming granulomas (Dales, 1978).
These older eggs were considered by Smith (1976) to have parthenogenetic potential and the possibility of facultative parthenogenesis
cannot be totally excluded.

According to Maynard Smith (1978), if sibling fertilisation
alternates with random outbreeding then the evolutionary stable sex
ratio is a little over 0.3. The lack of dispersal phases in adult
and larval N.diversicolor creates suitable conditions for such an
intensely inbreeding population. Genetic differences between locally
isolated populations have been proposed in adjacent estuaries (e.g.
Bryan, 1974) and clinal variation is suggested by two studies of
paragnath (buccal teeth) numbers in populations along a salinity
gradient (Barnes, 1978; Varialle, 1973).

These mechanisms distorting the sex ratio all stem directly from
atoky as an estuarine adaptation and none can be completely discounted
on present evidence. It is, of course, possible that the tertiary
sex ratio is not part of a strategy as such, but is imposed on the
worms by selective attack of disease, parasite or predator after the
period of parental care: then primary and secondary sex ratios would
be unaffected. Dales (1950) recorded ciliates feeding on spermatocytes in a 'very large percentage' of Thames worms and a virus is
known which only attacks the spermatocytes of male N.diversicolor
(Devauchelle & Durchon, 1973).

Whatever the reason for a preponderance of females, the population will benefit, provided enough males are available to fertilise
the eggs, since the rate of increase of the population is faster for
having more egg producers for a fixed density of adults.

DISPERSAL

Most animals have a dispersal phase in their life cycle. The benefits of dispersing to reach new territories must be set against the benefits of remaining safely within the preferred habitat. Nereids which brood or have benthic larvae will have limited dispersal, although swimming excursions at any stage in the life cycle may ensure distribution. For littoral animals, there may be no advantage in having a planktonic dispersal phase longer than six hours (Crisp, 1976) and brief periods of swimming have been observed in larvae of N.diversicolor and N.virens. Although fertilised eggs of N.japonica sink, the trochophores swim to the surface and descend again as 3 or 4 setiger larvae after a pelagic phase, measurable in days rather than hours (Inamori & Kurihara, 1979c).

It might be supposed as the population density increased, that a migratory urge would be stimulated through the greater interaction between individuals. In this case, a removal of surplus worms by emigration would still be compatible with locally inbreeding groups. Reise (1979a) found evidence for behavioural spacing, possibly resulting from contacts made during feeding excursions. Aggression, both inter- and intra-specific, is greater in species with epitokous breeding and declines to indifference in some atokous species (references in Evans, 1973) but Evans (1973) found immature N.diversicolor responsive, by attack or withdrawal, to contact with others so that the potential for behavioural regulation of population density exists even in the atokous species. This could occur at any stage in development.

Field evidence for migratory movement is available for N.virens: Blake (1979) found recolonisation of patches of cleared beach to normal density levels in two or three months but whether by encroachment from the edges or arrival of swimming worms was not ascertained. Dean (1978) observed immature N.virens swimming at night in winter months but Goerke (1979) found that less than 1% of worms, held in culture in the year before maturing, left their tubes.

Neither Brafield & Chapman (1967) nor Bass & Brafield (1972) found larvae of N.virens on the flats of Southend (Thames estuary) and onshore migration by juvenile worms was proposed. Snow & Marsden (1974) found larvae on the shore with adult N.virens and no evidence for a short swimming phase. Onshore migration of the 0+ class of N.diversicolor in the Tees estuary was described by Evans et al., (1979) although they allowed that these small worms could have been missed by the sampling method used. Chambers & Milne (1975) also failed to find juvenile worms. A fine mesh (around 200 microns) is needed to retain the smallest setigerous larvae and studies using sieves this fine have shown all sizes of worm to be present on the shore (Mettam, 1979; Ratcliffe, 1979).

Varialle (1973) found a separation in habitat between small
and large N.diversicolor, the division being at about 40 mm length,
which suggests a mobility of stocks. Indirect evidence for upstream
movement of 0+ class worms in the Severn estuary was obtained,
recruiting to a population (at Awre), but small worms removed by
surface scouring in spring at another Severn site (Lydney) did not
return over several months (Mettam, 1979).

Populations of both N.virens and N.diversicolor clearly do
disperse and probably can do so at any stage in the life cycle,
although the existence of a dispersive strategy remains to be
demonstrated.

LIFE CYCLES

Life cycles are subject to many strategic variations (Stearns,
1976). Reproduction is semelparous in the Nereidae (Olive & Clark,
1978). As a strategy to maximise the total number of offspring in
the lifetime of an individual, semelparity is appropriate for swarming
nereids which must suffer heavy losses to predators during their
pelagic nuptial excursion. The evolution of semelparity is an
irreversible path, according to C.S. Williams (quoted by Stearns,
1976).

The link between spawning and swarming requires tight synchrony
within the breeding population, but not necessarily between populations of a species within the same general area (review: Clarke,
1961). Once the link is broken, in atokous breeding, this demand
is relaxed and the vulnerable phase of pelagic life is avoided.
This would seem to encourage an iteroparous strategy in estuarine
nereids but none is known to have departed so far from family
tradition.

The option of investment in either somatic or reproductive
growth depends on the changing titre of a cerebral hormone, or
hormones, stimulating growth and inhibiting maturation (review:
Olive & Clark, 1978). Only recently the various effects of changing
hormone levels in N.diversicolor have been analysed simultaneously
(Al-Sharook & Golding, 1979), revealing a capacity for continued
growth after total release of oocyte inhibition. This accords with
field observations of continued growth during maturation and after
the spawning time (Mettam, 1979; Dales, 1950).

Nereis virens, N.succinea and N.diversicolor, in spite of their
very different reproductive modes, occupy overlapping habitats and
are potential competitors (Wolff, 1973). The large and potentially
predatory N.virens and the aggressive N.succinea may oust
N.diversicolor where they meet but the mechanism of competition is
not well studied. By whatever means, N.virens has displaced
N.diversicolor from shallow Danish waters since 1950 (Smith, 1977)

confining the latter to local areas of reduced salinity. This
observation suggests a hypothesis that each species has a different
strategy, compromising between physiological adaptability on the
one hand, and high growth rate, large size and associated fecundity
on the other, and that N.diversicolor, with extreme physiological
tolerance, has a slower growth rate even in environments where it
is not particularly stressed.

It is possible to explore this hypothesis to some degree,
although the foundation of fact is not as secure as it might be.
There appears to be regional variability in the life cycle of
N.diversicolor (references in Dales, 1950; Heip & Herman, 1979)
but recent information suggests that some of this may be more apparent
than real. Dales (1950; 1951) inferred a life cycle of one year
for N.diversicolor with a possible longevity of 18 months for
females that had failed to spawn in February. Severn estuary
populations from a wide range of salinities all spawned in May
following two years of somatic growth and a few of the 2+ group
lived on at least until September with the possibility of continued
spawning through the summer (Mettam, 1979). The two accounts are
compatible if a year group of worms is interpolated into the gap of
size classes which Dales (1950; 1951) did not sample. The time to
maturity in N.diversicolor is consistently 2 years from other sites
(Evans et al., 1979; Ratcliffe, 1979).

Differences in reproductive cycles do occur in other nereids in
different parts of their range. Nereis succinea, probably a summer
breeder in northern Europe, breeds throughout the year at Salton Sea,
California (Kuhl & Oglesby, 1979). Laeonereis culveri has a single
breeding season in Connecticut but a double one in Mexico and Central
America (Mazurkiewicz, 1975).

A peak of spawning in May seems to be a general feature of
N.diversicolor from sites with very difficult salinity and temperature
regimes (Heip & Herman, 1979; Evans et al., 1979; Ratcliffe, 1979).
Smith (1976; 1977) supposed a second spawning in autumn from the
presence of three size classes of worms in summer. This observation
is consistent with the two year cycle, single breeding peak, however.
A double breeding peak was supposed from the Ythan estuary, Scotland
(Chambers & Milne, 1975) but based on indirect evidence.

A strategy to maximise survival of offspring might coincide
reproduction with a time of optimal feeding conditions for the new
generation. If this is the reason for the May spawning period it
is not always a successful strategy in northern waters. Smith (1976)
found that unusually warm conditions at that time of the year could
'overripen' eggs and reduce fertility in intertidal populations
while cool subtidal populations matured more slowly and remained
fertilisable for longer. By breeding early in the year, worms were
not able to penetrate the low salinity waters of the inner Baltic.

Heip & Herman (1979) recorded continuous recruitment through the summer in a shallow pool. They calculated a linear growth increase of 0.05 mm per day. On this basis the largest worms in their samples, nearly 75 mm, would be of the 3+ year group, but their more accurate data fit (Heip & Herman, 1979) gives a maximum life of little over 3 years. Prolonged breeding may be associated with the unusual pool habitat. Nereis diversicolor that invaded an artificial tidal flat at the Texel laboratory showed waves of reproduction (de Wilde & Kuipers, 1977) and Bogucki (1953; 1962), bringing worms into culture in December, was able to hasten spawning and to rear a new generation to spawn in September. Natural spawning, he maintained, took place in April to May after one or two years of growth.

Brafield & Chapman (1967) concluded, from a study of the oocyte cycle and population weight frequency analysis, that N.virens at Southend typically spawns in April after two years of somatic growth. This interpretation matches the life cycle of N.virens with that of N.diversicolor, as previously described, in spite of the much greater size of the former species, but it has not gone unchallenged. Snow & Marsden (1974) proposed that N.virens from New Brunswick matured after 4 years (with a small proportion earlier). Snow & Marsden (1974) found larvae up to 6 segments and small worms from 73 - 90 segments onwards which, in summer, were taken to be 0+ group and 1+ group respectively: the growth rate of larval N.virens in culture was not faster than Dales (1950) found for N.diversicolor and the larger size of N.virens was taken as reflecting a longer life cycle.

Laboratory studies (Kay & Brafield, 1973; Goerke, 1979) support the faster growth and shorter life cycle of N.virens but culture conditions may substantially alter the natural cycle, as Bogucki (1962) found for N.diversicolor. Shumway (1979) compared both nereids collected at Anglesey, N. Wales and found that N.virens, in spite of the larger size of the specimens used, had a higher metabolic rate than N.diversicolor. Other comparisons can be made from feeding and growth rates measured experimentally.

FEEDING

Feeding in nereids has been very fully reviewed by Goerke (1971a). Several feeding modes may be used but no study of optimal foraging strategy has been made. Recent studies include those of Tsuchiya & Kurihara (1979), Jørgensen (1979), Reise (1979b).

Ingestion rates in N.succinea and N.diversicolor were compared by Veltishera & Karzinkin (1970) (quoted by Cammen, 1980) who found a gut contents turnover rate 2-3 x faster in N.diversicolor. Neuhoff (1979) measured consumption (C), assimilation (A) and production (P) in these two species at different temperatures and salinities for 28 day periods. Gross growth efficiency ($K_1 = P/C$)

values for N.diversicolor were about 2 x those of N.succinea at 15°C
for a standard size animal. Net growth efficiencies ($K_2 = P/A$)
indicate superior food conversion ability in N.virens and accord with
the rapid growth attributed to this species (Table 4 in Neuhoff,
1979). Part of the difference between N.diversicolor and N.succinea
lies in the demand for warmer conditions for feeding, and for breeding, in the latter (references in Neuhoff, 1979).

Inamori & Kurihara (1979a,b) showed the effects of environmental
variables, including density of worms, on food conversion efficiency
and growth rate of N.japonica. Laboratory experiments indicate a
direct effect of temperature on feeding e.g. Goerke (1979) found that
handling (swallowing) time was temperature dependent in N.virens and
that food was not swallowed below 4°C.

Low temperatures and salinities may account for the widely
observed cessation of growth in winter (e.g. Dales, 1950; 1951;
Chambers & Milne, 1975; Mettam, 1979; Ratcliffe, 1979; but not
Heip & Herman, 1979 for N.diversicolor) but feeding opportunity will
also vary seasonally. Numbers and biomass of N.diversicolor
increased by more than an order of magnitude in the better feeding
grounds towards the vicinity of an outfall in Kiel Bay (Anger, 1977).
The environmental determinants of feeding will vary from site to
site, yet the field evidence suggests some uniformity of growth
rates at least in N.diversicolor.

The seasonal cycle of growth could be environmentally determined
or an evolutionary strategy. Low levels of activity and poor
condition in winter could improve survivorship of individual worms
faced with intensive predation. In areas where bird predators may
take up to 90% of available N.diversicolor (Evans et al., 1979) the
speculation is not unreasonable. Nereis diversicolor, with a lower
metabolic rate and growth rate may be less vulnerable to predators
than N.virens and predators selecting the larger species as prey,
could keep N.virens from intertidal mudflats, releasing N.diversicolor
from competition. A biological interaction of this kind could inter-
act with differences in physiological tolerance.

PHYSIOLOGY AND REGULATION MECHANISMS

Differences in salinity tolerances and regulation mechanisms
of estuarine nereids have been reviewed many times (e.g. Oglesby,
1978)). In general, there is a good correspondence between an
osmoregulatory capacity and habitat. Theede et al. (1973) recorded
highest salinity limits for N.succinea below full strength sea water
but the same species occurs in the hypersaline Salton Sea, California
and it is capable of thriving in even higher salinities (Kuhl &
Oglesby, 1979).

One of the main advantages of colonising estuaries must be the

organic richness of the muds which provide an abundance of micro-
organisms, although these present the hazard of potential infection
against which worms must protect themselves (Dales, 1978). Organically rich sediment tend to be anaerobic and this can be a costly
stress to animals. The responses to low oxygen levels are different
in different nereids and correspond to the degree of hypoxia to which
they are exposed in nature (Theede et al., 1973). Schöttler, (1979)
measured end products of anaerobic metabolism in N.pelagic, N.virens
and N.diversicolor, a sequence of increasing tolerance, and measured
'stress' imposed by anaerobis before and after exposure. All three
species produced large amounts of lactate, which also accumulated
under aerobic conditions during muscular activity. N.diversicolor
switched to producing succinate, propionate and acetate, an
energetically "more convenient" pathway, under continued anaerobis.
N.pelagica had very little capacity to do this and N.virens was
intermediate.

SUMMARY

     Strategic compromises between rapid growth, large size and
fecundity on the one hand, and physiological adaptability on the
other, employed by different, competing species of nereid polychaete,
may partly explain their different distribution within estuaries.
The fact of atokous reproduction in Nereis diversicolor allows
speculation that the unbalanced sex ratio in this species is a
strategic adaptation but the evidence suggests that it is imposed
by outside influences.

REFERENCES

Al-Sharook, Z.M. & Golding, D.W. 1979. Patterns of cerebral endocrine
     activity in Nereis diversicolor Annelida Polychaeta. Marine
     Behaviour and Physiology, 6 (2): 95-104.
Anger, K. 1977. Benthic invertebrates as indicators of organic
     pollution in the western Baltic Sea. Internationale Revue der
     gesamten Hydrobiologie, 62 (2): 245-254.
Barnes, R.S.K. 1978. Variation in the paragnath number in Nereis
     diversicolor in relation to sediment type and salinity regime.
     Estuarine and Coastal Marine Science, 6: 275-283.
Bass, N.R. & Brafield, A.E. 1972. The life cycle of the polychaete
     Nereis virens. Journal of the marine biological Association
     of the United Kingdom, 52: 701-726
Blake, R.W. 1979. On the exploitation of a natural population of
     Nereis virens from the northeast coast of England, U.K.
     Estuarine and Coastal Marine Science 8 (2): 141-148.
Bogucki, M. 1953. Rozród i rozwój wieloszczeta Nereis diversicolor
     (O.F. Muller) w Baltyku. Polskie Archiwum Hydrobiologii 1(14):
     251-270.
Bogucki, M. 1962. Howdowla Nereis diversicolor O.F. Muller w
     warunkach laboratoryjnych. Przeglad Zoologiczny 6 (3): 232-234.

Brafield, A.E. & Chapman, G. 1967. Gametogenesis and breeding in
    a natural population of Nereis virens. Journal of the marine
    biological Association of the United Kingdom, 47: 619-627.
Bryan, G.W. 1974. Adaptation of an estuarine polychaete to sediments
    containing high concentrations of heavy metals. In, Pollution
    and physiology of Marine organisms (F.J. Vernberg & W.B.
    Vernberg, eds) pp. 123-135. Academic Press.
Cammen, L.M. 1980. A method for measuring the ingestion rate of
    deposit feeders and its use with the polychaete Nereis
    succinea. Estuaries, 3 (1): 55-60.
Chambers, M.R. & Milne, H. 1975. Life cycle and production of
    Nereis diversicolor O.F. Muller in the Ythan estuary, Scotland.
    Estuarine and Coastal Marine Science, 3: 133-144.
Clark, R.B. 1961. The origin and formation of the heteronereis.
    Biological Reviews, 36: 199-236.
Crisp, D.J. 1976. The role of the pelagic larva. In, Perspectives
    in Experimental Biology (P.S. Davies, ed) vol. 1, Zoology,
    pp. 145-155, Pergamon Press.
Dales, R.P. 1950. The reproduction and larvae development of Nereis
    diversicolor O.F. Muller. Journal of the Marine Biological
    Association of the United Kingdom, 29: 321-360.
Dales, R.P. 1951. An annual history of a population of Nereis
    diversicolor O.F. Muller. The Biological Bulletin, 101: 131-137.
Dales, R.P. 1978. Defence Mechanisms. In, Physiology of Annelida,
    (P.J. Mill, ed), pp. 479-507. Academic Press.
Dean, D. 1978. Migration of the sandworm Nereis virens during
    winter nights. Marine Biology (Berlin) 45 (2): 165-173.
Devauchelle, G. & Durchon, M. 1973. Sur la presence d'un virus,
    du type iridovirus dans les cellules males de Nereis
    diversicolor (O.F. Muller). Compte rendu hébdomadaire des
    séances de l'Académie des Sciences, Paris, 277 (4): 463-466.
Evans, P.R., Herdson, D.M., Knights, P.J., Pienkowski, M.W. 1979.
    Short-term effects of reclamation of part of Seal Sands,
    Teesmouth, England, U.K. on wintering waders and shelduck 1.
    Shorebird diets, invertebrate densities and the impact of
    predation on the invertebrates. Oecologia (Berlin), 41 (2):
    183-206.
Evans, S.M. 1973. A study of fighting reactions in some nereid
    polychaetes. Animal Behaviour, 21: 138-146.
Fisher, R.A. 1930. The genetical theory of natural selection,
    Clarendon Press, Oxford.
Goerke, H. 1971a. Die Ernährungsweise der Nereis-Arten (Polychaeta:
    Nereidea) der Deutschen Küsten. Veröffentlichungen des
    Instituts für Meeresforschung in Bremerhaven, 13: 1-50.
Goerke, H. 1971b. Nereis fucata (Polychaetae, Nereidae) als
    kommensale von Eupagurus bernhardus (Crustacea, Paguridae).
    Entwicklung einer Population und Verhalten der Art.
    Veröffentlichungen des Instituts für Meeresforschung in
    Bremerhaven, 13: 79-118.

Goerke, H. 1979. Nereis virens (Polychaeta) in marine pollution research: cul.ture methods and oral administration of a poly-chlorinated biphenyl. Veröffentlichungen des Instituts für Meeresforschu ng in Bremerhaven, 17 (2): 151-162.

Heip, C. & Herman, R. 1979. Production of Nereis diversicolor O.F. Muller (.Polychaeta) in a shallow brackish water pond. Estuarine and Coastal Marine Science, 8 (4): 297-305.

Inamori, Y . & Kurahara, Y. 1979a. Analysis of the environmental factc rs affecting the life of the brackish polychaeta, Neant ches japonica (Izuka) i. The effects of the environmental fact ors on survival and growth. The Bulletin of the Marine Biol ogical Station of Asamushi, Tohoku University, 16 (3): 87-1 .00.

Inamori, Y. & Kurihara, Y. 1979b. Analysis of the environmental fac tors affecting the life of the brackish polychaete, Nea nthes japonica (Izuka) ii. Thermal and salty conditions req uired for development and growth. The Bulletin of the Marine Bic ological Station of Asamushi, Tohoku University, 16 (3): 101 [-112.

Inamori , Y. & Kurihara, V. 1979c. Analysis of the environmental fa ctors affecting the life of the brackish polychaete, Neanthes j aponica (Izuki) iii. The effects of the environmental factors on fertilisation, cleavage and post-larval development. The l Bulletin of the Marine Biological Station of Asamushi, Tohoku U niversity, 16 (3): 113-121.

Inamori, Y. & Kurihara, Y. 1979d. Analysis of the environmental fa ctors affecting the life of the brackish polychaete Neanthes jap onica (Izuki) V. Thermal alternation as an environmental fac tor inducing sexual maturation. The Bulletin of the Marine Biol ogical Station of Asamushi, Tohoku University, 16 (3): 129- 139.

Jorgensen , N.O.G. 1979. Uptake of L valine and other amino acids by the polychaete Nereis virens. Marine Biology (Berlin), 52: 45-52.

Kay, D.G. & Brafield, A.E. 1973. The energy relations of the poly-chaet e Neanthes (Nereis) virens (Sars). Journal of Animal Ecolc gy, 42: 673-692.

Kuhl, D.L. & Oglesby, L.C. 1979. Reproduction and survival of the pilew orm Nereis succinea in higher Salton Sea salinities. The Biolo gical Bulletin, 157 (1): 153-162.

Maynard Smi th, J. 1978. The evolution of sex. Cambridge University Press.

Mazurkiewic z, M. 1975. Larval development and habits of Laeonereis culveri (Webster) (Polychaeta: Nereidae). The Biological Bulleti n, 14.9: 186-204.

Mettam, C. 1 979. Seasonal changes in populations of Nereis diversicolor O.F. Muller from the Severn Estuary, U.K. In, Cyclic pheromena in marine plants and animals (E. Naylor & R.G. Harknoll, eds) pp. 123-130. Pergamon Press.

Neuhoff, H. G. 1979. Influence of temperature and salinity on food conver sion and growth of different Nereis species (Polychaeta,

Annelida). Marine Ecology Progress Series, 1 (3): 255-262.

Oglesby, L.C. 1978. Salt and water balance. In: Physiology of Annelida (P.J. Mill, ed) pp. 555-659. Academic Press.

Olive, P.J.W. & Clark, R.B. 1978. Physiology of Reproduction. In: Physiology of Annelida, (P.J. Mill, ed) pp. 271-368, Academic Press.

Pianka, E.R. 1978. Evolutionary Ecology (2nd Edition). Harper & Row.

Reise, K. 1979a. Spatial configurations generated by motile benthic polychaetes. Helgoländer wissenchaftliche Meeresuntersuchungen, 32 (1-2): 55-72.

Reise, K. 1979b. Moderate predation on meiofauna by the macrobenthos of the Wadden Sea. Helgoländer wissenchaftliche Meeresunter-suchungen, 32: 453-465.

Rasmussen, E. 1956. Faunistic and biological notes on marine invertebrates iii. The reproduction and larval development of some polychaetes from Isefjord with some faunistic notes. Biologiske Meddelelser udgivet af Det Kongelige Danske videnskabernes selskab, 23 (1): 1-84.

Ratcliffe, P.J. 1979. An ecological study of the intertidal invertebrates of the Humber estuary. Ph.D. Thesis, University of Hull.

Schöttler, U. 1979. On the anaerobic metabolism of three species of Nereis (Annelida). Marine Ecology Progress Series, 1 (3): 249-254.

Shumway, S.E. 1979. The effects of body size, oxygen tension and mode of life on the oxygen uptake rates of polychaetes. Comparative Biochemistry and Physiology A. Comparative Physiology 64 (2): 273-278.

Smith, R.I. 1976. Further observations on the reproduction of Nereis diversicolor (Polychaeta) near Tvarminne, Finland and Kristeneberg, Sweden. Annales Zoologici Fennici, 13: 179-184.

Smith, R.I. 1977. Physiological and reproductive adaptations of Nereis diversicolor to life in the Baltic Sea and adjacent waters. In Essays on Polychaetous Annelids in memory of Dr. Olga Hartman (D.J. Reish & K. Fauchald eds). pp. 373-390. Special Publication of the Allan Hancock Foundation. University of Southern California.

Snow, D.R. & Marsden, J.R. 1974. Life cycle, weight and possible age distribution in a population of Nereis virens from New Brunswick. Journal of Natural History, 8: 513-527.

Stearns, S.C. 1976. Life history tactics: a review of the ideas. Quarterly Review of Biology, 51: 3-47.

Theede, H., Schandinn, J. & Saffé, F., 1973. Ecophysiological studies on four Nereis species in the Kiel Bay. Oikos 15 (Supplement) 246-252.

Tsuchiya, M. & Kurihara, Y. 1979. The feeding habits and food sources of the deposit feeding polychaete, Neanthes japonica (Izuka). Journal of experimental marine Biology and Ecology, 36: 79-89.

Varriale, A.C. 1973. Characteristiques morphologiques et ecologiques d'une population de Nereis diversicolor des eux saumatres de

Livourne, Cahiers de Biologie Marine, 14: 1-10.

Wilde, P.A.W.J. de & Kuipers, B.R. 1977. A large indoor tidal mud flat ecosystem. Helgoländer wissenchaftliche Meeresunter-suchungen, 30 (1-4): 334-342.

Wolff, W.J. 1973. The estuary as a habitat. An analysis of data on the soft bottom macrofauna of the estuarine area of the rivers Rhine, Meuse, and Sheldt, Zoologische Verhandelingen uitgegeven door het rijksmuseum van naturlijke historie te Leiden, 126: 1-242.

BEHAVIOURAL ACTIVITIES AND ECOLOGICAL STRATEGIES IN THE INTERTIDAL

GASTROPOD HYDROBIA ULVAE

R.S.K. Barnes

Department of Zoology
University of Cambridge
Cambridge, CB2 3EJ

INTRODUCTION

   This paper interweaves two cautionary tales, one emphasising
our lack of good, descriptive information, the other suggesting that
too readily do we interpret behavioural activities in terms of
plausibly argued 'strategies' which are more satisfying aesthetically
than they are in accord with field observations.

   If a European estuarine zoologist be asked "Of which estuarine
invertebrate's field biology do we have the most complete picture?",
there is a high probability that the small, intertidal prosobranch
Hydrobia ulvae (Pennant) would be nominated. Certainly it would be
regarded as being amongst the five best known and most thoroughly
studied members of the macrofauna, probably largely as a result of
its ubiquity and its occurrence in densities which may, on occasion,
exceed $100,000m^{-2}$. The activity patterns of H.ulvae and their
significance have been summarized by Newell (1979, pp 277-8), based
largely on his own earlier work, thus:

"when the tide is out the animals can be seen crawling about ... .
After a time, the animals burrow and remain beneath the surface of
the substratum actively feeding on material near the surface until
just before the incoming tide reaches them. Then most of the
animals resurface and crawl up the ripple marks to launch themselves
upside down on the surface film of the water by means of a mucous
raft. At this stage the animals actively feed on material trapped
in the raft, and continuously secrete mucus from the foot. The in-
flowing tide then carries the snails shorewards and later strands
them on the mud where the deposits are fine and the slope shallow.
("after a time which varies in populations from different localities,

79

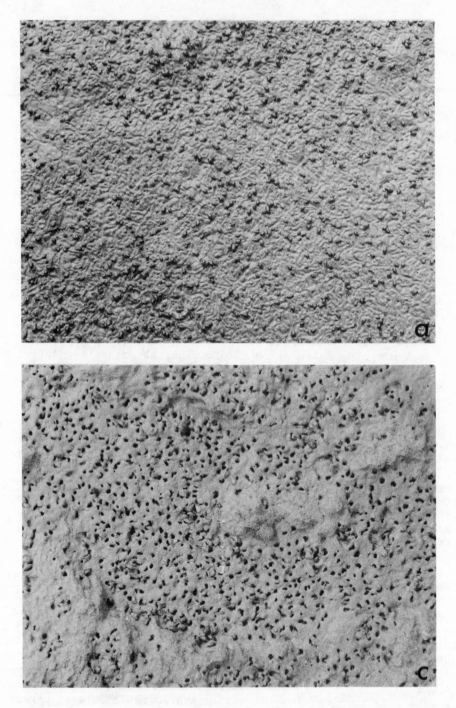

Figure 1. <u>Hydrobia</u> <u>ulvae</u>: (a) crawling; (b) climbing; (c) buried;
(d) floating.

they sink to the bottom ... and regain their characteristic intertidal position ... the animals do this where there is a sudden
lessening of the slope of the shore" (Newell, 1962, pp. 72, 74 & 73
respectively)). They begin browsing again and the cycle of behaviour
is completed. This sequence ... allows the animal to utilise the
whole of the tidal cycle for feeding activities".

Newell also considered that this behavioural cycle accounted for the
lack of sediment preferences displayed by the snail in his tests:

"because ... the sinking methods ... are adequate to ensure that the
majority of the floating animals will be deposited in a region where
the gradient is slight which is in turn correlated with a fine grade
of deposits. It therefore becomes unnecessary for the animals to be
able to search for a favourable deposit in which to burrow". (Newell,
1962, p. 63).

Anderson (1971) studied the climbing and floating activities of
H.ulvae in detail and concluded that the advantage of climbing (in
this case of whisky bottles set into the sediment) lay in the subsequent flotation from the climbed structures, which in turn was
suggested to achieve dispersal of the snails to other areas of the
estuary (the Ythan). Earlier, Muus (1967), amongst others, had also
concluded that: "the tendency of the Hydrobiae to crawl upwards
under unfavourable conditions provides them with especially good
possibilities for passive dispersal to other biotopes" (p. 154).
Levinton (1979) has argued similarly in respect of Hydrobia ventrosa
Montagu.

These series of strategies adapting the feeding of H.ulvae to
the intertidal regime and its dispersal to the prevailing estuarine
conditions have become classics, and the story has been repeated in
several textbooks. This paper reviews evidence presented by the
author (Barnes & Greenwood, 1978; Barnes, 1979, 1981a, 1981b &
unpublished) and others (e.g. Vader, 1964; Schäfer, 1972; Little
& Nix 1976) which suggests that although H.ulvae does indeed crawl,
burrow, climb and float (Fig. 1), the accepted interpretation
summarized above is seriously in error. Most importantly, these
activities do not form part of a behavioural cycle, and climbing
and floating, when they occur, are not strategies for dispersal.

CLIMBING

If climbing is but one phase of a regular cycle of behavioural
activities, then all Hydrobia able to climb (i.e. which encounter
some vertical element in their immediate vicinity) should do so.
Alternatively, if climbing has evolved as a mechanism to achieve
flotation, which is itself the dispersal strategy, then one can
predict that climbing activity would be most marked in centres of
high population density and/or under the "unfavourable conditions"

discussed by Muus (1967). (Anderson, 1971, has shown that climbing/ floating are not related to the reproductive cycle). A further prediction is that all snails which have climbed vertical structures would there await the opportunity to gain the water surface.

None of these predictions was borne out by the field and laboratory studies of Barnes (1981a, 1981b). Further, in many hours of observation, no snail was ever observed to climb in advance of being covered by the rising tide and then to launch itself on the surface film when the latter reached it; climbing only occurred after being covered by the tide, either when the animal was under water or whilst the sediment or other surface was still wet immediately after the ebb (see also Little & Nix, 1976). When the number of snails which had climbed Salicornia plants growing from a mud-flat was assessed in relation to the population density of Hydrobia in the $0.01m^{-2}$ surrounding each plant (with total densities of snails ranging from 17,700 to $137,300m^{-2}$), the proportion which had climbed varied between 0.4% and 38.2% without any correlation existing between numbers having climbed and total population density (Spearman Rank Correlation 0.27; P>0.05) (Barnes, 1981a). When climbing behaviour was assessed under unfavourable conditions (41% sea water) and compared to favourable ones (100% sea water), more than one hundred times more snails climbed under the favourable regime (Barnes, 1981b).

Nevertheless, a marked and consistent pattern of climbing activity was observed, in both field and laboratory, which varied in relation to (a) period of tidal cover and (b) position of the covering tide in the spring tidal sequence (Fig. 2). Two features of this pattern are particularly relevant here. First, having climbed up vertical stakes on being covered by the tide, many snails climbed back down again and on to the sediment before the ebb. Secondly, climbing activity (as measured by the number of snails having climbed after 30 minutes of tidal cover) was inversely corre- lated, very strongly, with the cumulative period of cover (and hence - see below - of feeding opportunity) by the previous six or seven covering tides (Spearman Rank Correlation 0.95; P << 0.001).

These results suggested a relationship between climbing and feeding, and a lack of relationship between the observed pattern of climbing and that which would be expected if climbing was related to dispersal. These were further tested by comparing climbing activity on vertical structures (a) bearing abundant and suitable food and (b) without any edible material. A feeding hypothesis would predict greater numbers on the food-bearing structures; a dispersal hypoth- esis would predict greater numbers on those without. The results (Barnes, 1981b) showed that the food-bearing stakes always bore more climbed snails (Mann-Whitney U. zero; P << 0.001). This leads to the conclusion that climbing can be regarded as normal crawling/browsing behaviour which happens to be carried out in a non-horizontal plane

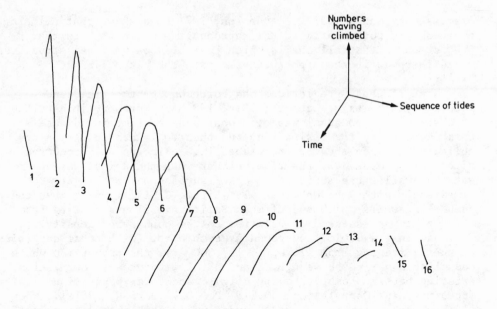

Figure 2. Three-dimensional representation of the climbing activity
of Hydrobia ulvae during an artificial spring-tide sequence (16 tides
ranging from 30 minutes cover – tides 1 & 16 – to 210 minutes cover
– tides 8 & 9). The N axis displays the mean number of snails up
three vertical stakes at different times within each period of tidal
cover (mean max. per stake 128); the t axis displays duration of
tidal cover; and the 16 tides are equally spaced along the third
axis. (After Barnes, 1981a, 1981b).

when that plane potentially bears suitable algal food, and that those
Hydrobia observed during periods of low tide to have climbed are
simply those which, having climbed during the previous high tide,
were marooned there on the ebb. In other words, climbing is not a
distinct behavioural strategy at all, any more, for example, than
it is in many prosobranchs on rocky shores.

Climbing (and crawling activity generally) were also observed
to be more prevalent during darkness than in daylight (Wilcoxon
Matched-pairs, Signed-ranks Test, $P < 0.005$). This can be interpreted
as an adaptation minimising predation by, for example, visually hunting predators such as plovers (Barnes, 1981b).

FLOATING/DISPERSAL

Climbing, then, does not seem to show the nature or pattern
which one would expect a flotation-preliminary to have. Yet, climbing would appear to be a necessary prerequisite to flotation. This

apparent paradox can be resolved by considering the nature of the
act of flotation itself. Is flotation achieved passively and accid-
entally, or is it an active process forming part of a behavioural
cycle? All the published evidence (e.g. Vader, 1964; Anderson,
1971; Little and Nix, 1976; Barnes, 1981a) indicates that only a
minute percentage of the intertidal populations of H.ulvae float on
any given tide. Of the 25 stations worked by Little and Nix, for
example, in only two was the percentage greater than 0.5% and in
fourteen it was zero. Even Anderson, who supported the climbing-
floating-dispersal hypothesis, only recorded a maximum of 200 snails
in an 80m-length water-surface sample taken over sediment supporting
an average of 46,000 H.ulvae m^{-2}.

The observations presented by these authors suggest that those
snails which float have done so by the following, entirely passive
procedure. Some H.ulvae are marooned up vertical structures on the
shore by the tidal ebb; as the surface beneath them dries, they
attach themselves to that surface each by a cord of mucus, close the
operculum and enter a phase of inactivity; during the remaining low
tide period, the snail shells and their associated mucus dry and, by
virtue of air bubbles and the dried mucus, become positively buoyant
with respect to sea water; and when finally the next high tide covers
them, some are floated passively from their erstwhile support.

The fate of inactive snails up vertical structures on being
covered by the tide was investigated by Barnes (1981a). When engulfed
by tidal water, one of three events occurred: (a) the snails resumed
their crawling activity on the structure; (b) they were dislodged and
sank to its base; or (c) they were floated from it. Under laboratory
conditions designed to maximise the opportunities for successful
floating, the respective percentages suffering these three fates were:
(a) 19.5, (b) 44.2, and (c) 36.4. When the water surface encroaching
on the snails was rippled, the values became 2.5%, 87.1% and 10.4%;
and even fewer floaters, percentages approximating those found by
Little and Nix (1976), were seen in the field from natural stands of
Salicornia. Of those that were floated from their support, the
majority fell to the substratum within only a few centimetres; and
several seconds, and in some cases minutes, elapsed before any of
the remaining snails emerged from their shells and extended the foot
along the surface film. Flotation, therefore, appears to be a rare,
accidental consequence of being marooned after climbing. It has no
active component and can hardly be considered an 'adaptive strategy'.

It is still.a common observation, however, that a few snails do
float for large distances, especially in extremely calm weather and
from isolated vertical structures. The extent to which this is of
significance with respect to dispersal will depend on the efficacy
of the other mechanisms for relatively long-distance dispersal available to H.ulvae. Here it is doubtless significant that H.ulvae is
the only western European hydrobiid with a planktonic larval stage.

Moreover, large numbers of this snail may also be observed being
swept or rolled along the surface of the sediment by flooding tides:
an (unpublished) sample count at Scolt Head Island, Norfolk, indicated
that the majority of the surface population of H.ulvae were moved
distances to be measured in metres by being rolled along the substratum. Indeed, unpublished observations on the north Norfolk
coast show that high spring tides roll and ultimately strand vast
numbers of Hydrobia to form a wide strand line, many individuals
deep and broad, at EHWS. This is a far from adaptive consequence
of this form of dispersal: in hot weather, the accumulations of
snails soon begin to rot. These two dispersal mechanisms, one acting on the larvae and the other on the adults, together with the
abilities of the adult snails to crawl quite large distances, clearly
reduce the need for flotation to serve as such, and except in rare
circumstances flotation probably is of little import.

BURIAL

    With the removal of climbing and floating from the standard
behavioural cycle, little of any 'cycle' remains, and even the
regularity with which burial during periods of exposure at low tide
occurs has been questioned (Little and Nix, 1976; Barnes, 1981a).
Newell (e.g. 1962) interpreted the burial phase as a strategy for
exploiting the subsurface deposits for food, and he portrayed the
animal as then lying with the aperture of the shell towards the
surface and the head and foot extended, feeding.

    H.ulvae does normally bury itself in the sediment as soon as
the substratum beings to dry, but examination of individual snails
which had so burrowed disclosed that they had all retreated into
their shells and closed the operculum (unpublished observations).
Schäfer (1972) found that most buried snails were not in a feeding
position but were lying with the shell aperture at the bottom of
the pit. These observations, and the response of Hydrobia up vertical
structures to drying of their substratum, suggest that H.ulvae burrows into the sediment to avoid desiccation (and probably also the
attention of surface feeding predators): data supporting the subsurface feeding hypothesis have not been forthcoming. It may well
be relevant here that Hydrobia in permanently flooded salt-pans,
pools and creek basins show a greatly lessened tendency to burial
(unpublished observations).

HABITAT SELECTION

    Finally, there is the question of the mechanism by which
H.ulvae aggregates in favourable areas, as the floating/sinking
'strategy' appears no longer to be available. Contrary to earlier,
albeit very preliminary, findings (Newell, 1962), this species is
now known to display very marked abilities to select behaviourally
preferred sediment types, although it is capable of inhabiting a

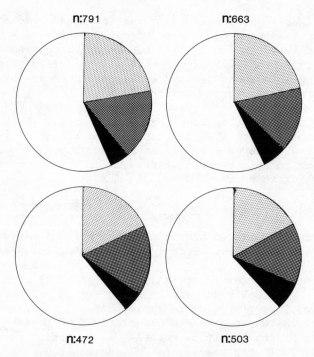

Figure 3. Pie-diagrams displaying the preference (four replicates)
of Hydrobia ulvae for different sediments. The species occurs
naturally in three of the sediments (not in that from the exposed
sandy creek); the experimental animals were taken from the sheltered sandy creek. Open quadrants; soft mud (mean particle diameter 30 μm, chlorophyll content 49 μg g^{-1} dry wt). Light stipple;
sheltered creek sand (m.p.d. 247 μm, 7 μg g^{-1} Chl.). Heavy stipple;
salt-pan sediment (mixture of c. 73% sand (m.p.d. 270 μm) and c.
27% silt (m.p.d. 16 μm), 11 μg g^{-1} Chl.). Black; exposed creek
sand (m.p.d. 264 μm, 14 μg g^{-1} Chl.). Overall χ^2 1627; P < 0.001.
(From data in Barnes and Greenwood, 1978; and unpublished).

wide range of substrata when these are the only ones available
(Barnes and Greenwood, 1978; Fig. 3). The preference itself may
be to the algal content of the sediment rather than to any aspect
of its physical nature (Coles, 1979). In these respects it is by
no means an atypical intertidal species (Meadows and Campbell, 1972)
and, as in other estuarine invertebrates (see, e.g., Ono, 1962),
its preference can be modified dependent on the population density
already supported by the substratum preferred. Thus the proportion
of snails choosing a less attractive alternative increases as the
total population density increases, although the generally-preferred
sediment type always supports the majority of the tested individuals

(Hammond and Weale, unpublished, cited in Barnes, 1981a). The very
wide range of substrata acceptable to this species as a whole may
have resulted in variation of the preferences displayed by local
populations: Barnes (1979) found that a small proportion of a population living in a sandy habitat apparently preferred that sand to
the more generally favoured finer sediments.

Therefore, during the crawling/browsing phase which occurs
when the animals are covered by the tide and for some time after
its ebb, H.ulvae is provided with an opportunity, and possesses
the mechanism, to select particularly suitable areas of sediment
on or over which to browse.

THE INTERTIDAL ACTIVITY OF H.ULVAE: A NEW SYNTHESIS

The recent studies of Hydrobia ulvae reviewed above present
a consistent picture which differs in many important areas from
our earlier ideas of this species. Elements of this picture are
(a) that climbing is merely normal feeding activity carried out in
an oblique or vertical plane and is not a strategy to facilitate
dispersal, (b) that flotation is an accidental, passive event to
which only a minute percentage of populations of H.ulvae are subject and that it can hardly be regarded as essential for dispersal,
(c) much dispersal will result from the possession of a planktonic
larval stage and from the adults being rolled across the sediment
surface by tidal water movements, (d) that burial is more plausibly
interpreted as a mechanism for avoiding desiccation, and possibly
predators, than as an alternative feeding strategy, and (e) that
the animal can select which micro-habitat to occupy during the
crawling phase. The snails do not surface from burial until after
the tide has covered them; and a mucous raft is not involved with
the initial act of flotation. In summary: the interpretation of
much of what we thought we knew of the field biology of H.ulvae was
in error (of how many other species is this also true?); and a
complex cycle of intertidal strategies adapting this gastropod to
a sequential series of available habitats does not appear to occur,
neither is it in any sense necessary.

Instead, the pattern of intertidal activity of H.ulvae appears
simply to be based on the presence or absence of tidal water cover.
On being covered by the tide, the snails crawl from burial to the
sediment surface and then continue crawling and feeding on any suitable substratum, whether vertically or horizontally oriented. This
activity continues for some time after the ebb, until the substratum
beings to dry. If that substratum is a horizontal, sedimentary one,
the snails bury themselves beneath its surface and remain quiescent
until once more covered by the tide. If the substratum is a hard
one into which burrowing is impossible, the snails retreat into
their shells, having, if marooned up a vertical structure, attached
themselves by a cord of mucus. In the latter case, on being covered

by the tide the snails are frequently displaced to the base of the
object climbed and there they resume crawling activity, or they may
begin again to crawl over the object itself, or they may be floated
off their support only to plummet to the substratum after travelling
a short distance. A few snails may float well away, and remain
floating for a considerable time before being 'sunk' by wave action,
collision with other objects, etc. During wind and rain, many (up
to 100%, Barnes, 1981a) of the snails marooned up vegetation, etc.
may be displaced back to the sediment during the low tide period
itself, before the flow of the tide reaches them.

These relatively simple activities are likely to adapt Hydrobia
to the fluctuating and often unpredictable regime of the estuarine
littoral much better than would any more complex, rigid cycle trigg-
ered by a series of responses to light and gravity and requiring
timing mechanisms which must be reset dependent on the animal's
position on the shore. Perhaps I may conclude by returning to the
two cautionary notes mentioned at the beginning of this paper. We
will not understand the ecology of estuarine animals, much less the
whole estuarine system, until more observations are made of the
field behaviour and daily lives of the species involved. We must
also exercise great caution when interpreting what appears to us to
be a distinct and discrete activity pattern in terms of an adaptive
strategy evolved to achieve some precise and specific end; a man
falling off a ladder may be distinctive but the act is probably
rarely strategic!

REFERENCES

Anderson, A. 1971. Intertidal activity, breeding and the floating
    habit of Hydrobia ulvae in the Ythan estuary. J. mar. biol.
    Ass. U.K., 51: 423-437.
Barnes, R.S.K. 1979. Intrapopulation variation in Hydrobia sediment
    preferences. Estuar. coast. mar. Sci., 9: 231-234.
Barnes, R.S.K. 1981a. An experimental study of the pattern and
    significance of the climbing behaviour of Hydrobia ulvae
    (Pennant). J. mar. biol. Ass. U.K., in press.
Barnes, R.S.K. 1981b. Factors affecting climbing in the coastal
    gastropod Hydrobia ulvae (Pennant). J. mar. biol. Ass. U.K.,
    in press.
Barnes, R.S.K. and Greenwood, J.G. 1978. The response of the inter-
    tidal gastropod Hydrobia ulvae (Pennant) to sediments of
    differing particle size. J. exp. mar. Biol. Ecol., 31: 43-54.
Coles, S.M. 1979. Benthic microalgal populations on intertidal
    sediments and their role as precursors to salt marsh develop-
    ment. In: Jefferies, R.L. & Davy, A.J. (Ed.), Ecological
    Processes in Coastal Environments pp. 25-42, Blackwell, Oxford.
Levinton, J.S. 1979. The effect of density upon deposit-feeding
    populations: movement, feeding and floating of Hydrobia
    ventrosa Montagu (Gastropoda: Prosobranchia). Oecologia, Berl.,

43: 27-39.

Little, C. and Nix, W. 1976. The burrowing and floating behaviour of the gastropod Hydrobia ulvae. Estuar. coast. mar. Sci., 4: 537-544.

Meadows, P.S. and Campbell, J.I. 1972. Habitat selection by aquatic invertebrates. Adv. mar. Biol., 10: 271-382.

Muus, B.J. 1967. The fauna of Danish estuaries and lagoons. Medd. Danm. fisk.- og Havunders. (N.S.), 5(1): 1-316.

Newell, R.C. 1962. Behavioural aspects of the ecology of Peringia (= Hydrobia) ulvae (Pennant). Proc. zool. Soc. Lond., 138: 49-75.

Newell, R.C. 1979. Biology of Intertidal Animals, 3rd Edn. Marine Ecological Surveys; Faversham.

Ono, Y. 1962. On the habitat preference of ocypoid crabs. Mem. Fac. Sci., Kyushu Univ. (E), 3: 143-163.

Schäfer, W. 1972. Ecology and Palaeoecology of Marine Environments. Oliver & Boyd, Edinburgh.

Vader, W.J.M. 1964. A preliminary investigation into the reactions of the infauna of the tidal flats to tidal fluctuations in water level. Neth. J. Sea Res., 2: 189-222.

THE SURVIVAL OF MACOMA BALTHICA (L.) IN MOBILE SEDIMENTS

P.J. Ratcliffe[1], N.V. Jones[2] and N.J. Walters[1]

[1] School of Science, Hull College of Higher Education
Cottingham Road, Hull
[2] Zoology Department, University of Hull, Hull, HU6 7RX

INTRODUCTION

The small lamellibranch Macoma balthica (L.) has been extensively studied in recent years. Many authors have investigated the spatial distribution pattern of Macoma and its growth and production, among them Anderson (1972), Beukema et al. (1977), Chambers and Milne (1975), McLusky and Allan (1976), Reading (1979) and Tunnicliffe and Risk (1977). However, causal relationships have seldom been demonstrated, and gaps have emerged in our knowledge of the factors influencing the distribution and mortality of the animal, particularly during its early life.

Whilst deposit feeding of Macoma has been studied in detail (Gilbert 1975), the effects of this activity upon the sediment and on young Macoma have not been considered. The high population densities of Macoma in the lower Humber afforded an opportunity to include such effects among potential mortality factors which could be tested under laboratory conditions.

The work here formed part of a seven year study of the distribution, abundance and production of intertidal invertebrates in the lower Humber estuary (Ratcliffe, 1979 and 1980).

SITE DESCRIPTION

Skeffling lies in the middle of the northern Humber shore bay known as Spurn bight (Figure 1). At this point, the estuary is over 11km wide at high tide. A Spartina salt marsh extends from just below M.H.W.S. tide level down the beach for approximately 80m. Further down the shore, no permanent rooted vegetation occurs, and

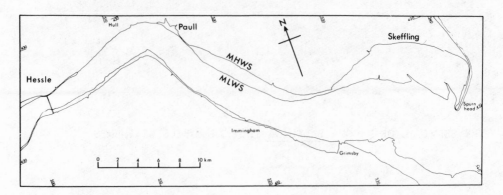

Figure 1. The middle and lower Humber estuary showing the position
of the Skeffling sampling area.

intertidal mud and sand flats extend for about 3km southwards.

The area is characterised by a high degree of exposure to wind,
with the prevailing wind being southerly and, therefore, directly
on-shore. The invertebrate macro-fauna of the intertidal zone contains many of the elements of the classical Macoma community (Thorson,
1957). Macoma balthica, Nereis diversicolor, Hydrobia ulvae, Retusa
obtusa, Cerastoderma edule, Nephtys hombergi, Mya arenaria and
Arenicola marina are all present in the middle and upper shores.
Macoma balthica is the most abundant bivalve species.

METHODS

The relative heights of 30 points situated at 10m intervals
running along a transect down the beach from M.H.W.N. tide level,
were determined using a levelling table. Direct observations of the
period of tidal immersion of points along this transect were also
made under both spring and neap tide conditions. The positions of
four primary sampling sites along this transect are indicated in
Figure 2.

Changes in the height of the sediment surface were monitored at
points along the transect by measuring the depth to which glass plates
buried beneath the sediment had been covered. The positions of the
plates were known in relation to marker stakes but no structure
affected the sediment surface immediately above the plates.

Samples of sediment for faunal analysis, particle size determination and organic content analysis were taken using plastic tubes.
Shell lengths of Macoma were measured to the nearest 0.5 mm and the
ages of the animals determined by analysis of length/frequency data
and checked by counting annual growth rings when these were visible.

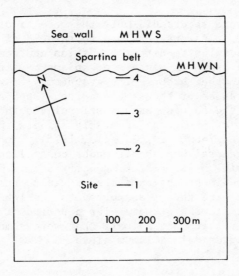

Figure 2. The Skeffling sampling area showing the positions of the transect sites.

Further details of the techniques used in sediment and faunal analysis are given in Ratcliffe (1979).

Continuous records of sediment temperatures were made using a multi-channel pen recorder and records of air temperature and wind speed and direction were made available by the Geography Department of the University of Hull.

Animals were maintained in their native sediment in the laboratory under natural illumination and photoperiod in clean aerated water at a constant salinity of 24‰ and a temperature that varied little from ambient and ranged daily between 6°C and 14°C. Animals for use in experiments were subjected to holding conditions for no more than 48 hours before the experiment started.

The high temperature tolerance of animals was tested by exposing groups of 10 animals in clean unaerated seawater in petri dishes to high temperatures in an incubator for periods of 6 hours. This short period was chosen to correspond with the maximum period that the animals might experience high temperatures whilst not covered by water. Death was assumed to have occurred if an animal showed no discernible movement and did not respond to mechanical stimulation.

Predation experiments were carried out by transferring predator and prey from their native sediment to small dishes containing sediment free of other macrofaunal species and covered by seawater. Predator and prey were left together for 96 hours, after which time numbers of both were recorded. Only the results of tests in which all the potential predators survived to the end of the test period were used. In addition to direct predation and competition, the activities of one group of benthic organisms may indirectly influence the survival of another group. The groups may be represented by different size classes of the same species. Such an interaction most often takes the form of an amensal relationship in which the feeding behaviour of one group adversely affects the survival of another (Rhoads and Young, 1970; Eagle, 1975). Apparatus was designed to test this general hypothesis under controlled laboratory conditions. Aluminium cartridges, 20 cm x 5 cm and 10 cm deep were pushed into the sediment and excavated so as to enclose intact cores of sediment which were then transported to the laboratory where they were subjected to flowing seawater for a period of an hour. Any large sediment particles and animals that were removed by this treatment were captured in a 125μm sieve.

RESULTS

Physical Features of the Skeffling Sites

The main feature of Figure 3 which shows a profile of the beach in the study area, is the increase in slope above site 3. Although not surveyed in detail, the beach below site 1 sloped even more gradually than between sites 1 and 3 for more than 1km.

Periods of immersion per tidal cycle ranged from 4 to 5 hours on neap and spring tides respectively at site 1, to 3 hours on a spring tide and 1 hour on a neap tide at site 4. Drying and cracking of the sediment surface occurred occasionally at times of neap tides during the summer at site 4.

Figure 4 reflects the instability of the sediments over a 27 month period at three points down the shore. A regular annual oscillation in sediment height of about 8cm is apparent at the uppermost site. Accretion occurred during early spring and erosion during autumn. Associated with this cycle, were changes in the appearance and consistency of the sediment surface layers from smooth and fluid during periods of maximum accretion, to rough and compacted during autumn and winter. Lower down the shore, annual oscillations were largely obscured by a steady increase in sediment level of about 8cm over the 27 month period, suggesting that the Skeffling area, at the time, was one of accretion.

Sediment particle size increased down the shore (Figure 5) whilst sediment organic content, which was closely correlated with the inverse of median particle diameter ($r = 0.812$, $p < 0.001$), decreased. Compared with changes in sediment type down the shore, temporal changes at any one site were much smaller, and, unlike changes in sediment height, showed no clearly defined seasonal changes. The

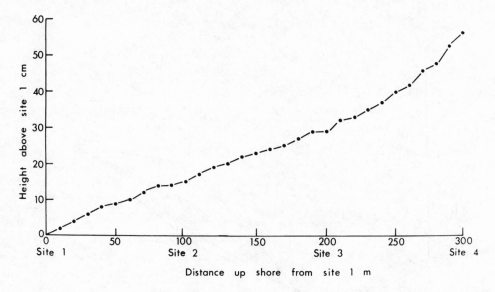

Figure 3.   The height of the beach along the   Skeffling transect.

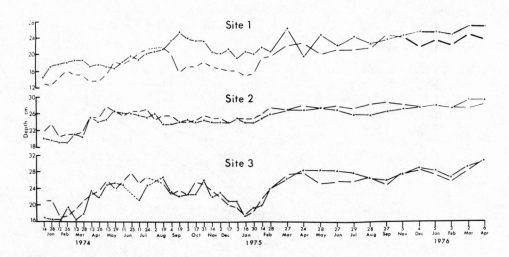

Figure 4.   The depths of plates below the sediment surface at three
Skeffling transect sites.

annual cycle of sediment height changes was not, therefore, associated with changes in sediment type.

Between January 1974 and April 1976 minimum temperatures at a

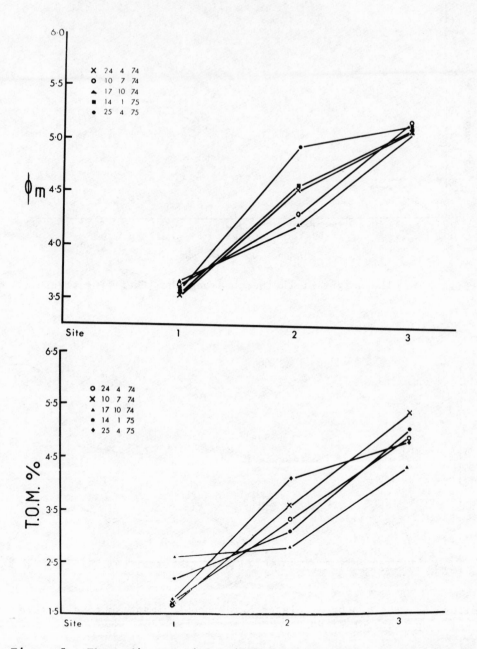

Figure 5. The median particle diameters and total organic matter contents of samples of sediment taken at three Skeffling transect sites.

depth of 10cm in the sediment fell to -2°C and on occasions ice was
observed on the Spartina marsh. Maximum sediment temperature in
1974 was 23.5°C and in 1975, 30.0°C. Figure 6 shows that the temperature of the exposed sediment surface followed air temperature
changes with only a few minutes lag. Temperatures deeper in the
sediment lagged further behind air temperatures. Even at a depth
of 6cm, however, responses to changes in air temperature occurred
within an hour. Recorded sediment temperature was always within
4.5°C of the air temperature and did not rise more than 0.5°C above
air temperature.

The hottest summer on record in recent years was 1976 when a
maximum air temperature of 31°C was recorded on 26.6.76 at a meteorological station near Skeffling. On this day, low tide was at noon
and the upper shore was exposed to the sun through the middle of the
day. A sediment temperature of 32°C may, therefore, have been
reached.

Temperature Tolerance of M.balthica Spat

Throughout the tests of high temperature tolerance, the results
of which are shown in Table 1, no mortality of Macoma spat was
observed. Even after enduring a temperature of 39°C for 6 hours,

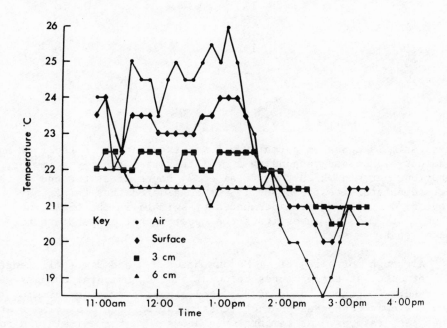

Figure 6. The temperature of sediment and air at site 3 at
Skeffling on 26.7.79.

Table 1. The Tolerance of Groups of 10 Macoma balthica Spat and 1+
          Animals to High Temperatures for a Period of 6 Hours

Temperature °C	Number of Macoma spat surviving 6 hours	Number surviving 24 hours after return to room temperature
20.0	10	10
24.0	10	10
27.0	10	10
30.0	10	10
32.5	10	10
35.5	10	10
37.5	10	10
39.0	10	10

Temperature °C	Number of 1+ Macoma surviving 6 hours	Number surviving 24 hours after return to room temperature
20.0	10	10
24.0	10	10
27.0	10	10
30.0	10	10
32.5	10	10
35.5	10	10
37.5	8	8
39.0	7	7

these animals appeared normally active, extending both the foot and
inhalent siphon and responding immediately to a tactile stimulus, a
behaviour pattern which they retained 24 hours after the end of the
high temperature test. Since it was considered that temperatures
above 39°C would not be experienced by animals in the field, no
attempt was made to investigate the upper lethal temperature for
Macoma spat.

     Although mortality of 1+ Macoma between 6 and 9mm shell length
did not occur at temperatures below 37.5°C, behavioural changes were
observed. At temperatures of 30°C and above, surviving animals
showed slower responses to tactile stimulation than did animals at
lower temperatures. At temperatures above 30°C the animals did not
retract the foot fully when the shell was touched, and those that
did survive at 37.5°C were also unable to retract the siphons fully.
After 24 hours at a lower temperature, however, the responses of

these animals were normal. Even at 39°C, the mortality recorded·
was only 30%.

Field Distribution of M.balthica

During the summer of 1979 densities of all sizes of <u>Macoma</u> were
monitored along the transect. Figure 7 shows that whilst the
density of animals other than spat remained relatively constant
during the period of study, the density of spat after spatfall at
the end of June, decreased lower down the shore but increased at
site 3. Above site 3, only small numbers of <u>Macoma</u> of any size were
found.

The change in the distribution of the spat did not take place
immediately after spatfall and is, therefore, interpreted as being
the result of shoreward transport of spat rather than heavier initial
settlement higher up the shore. The <u>Macoma</u> population at site 3 had
particularly large numbers of animals belonging to the 1+ age group,
suggesting that these animals had been transported to that area during
the previous 12 months.

Between September 1973 and April 1976, the total <u>Macoma</u> population at site 3, the area of maximum density, declined from about
40,000 m^{-2} to between 5 and 10,000 m^{-2}. This decline can be attributed to the mortality of the very dense 1973 settlement which
accounted for up to 90% of the total population numbers during
autumn 1973. A similar population age structure has been reported
by Chambers and Milne (1975). Settlement in 1974 and 1975 was very
much less successful, whereas 1976 was again a year of high recruitment and total densities reached 30,000 m^{-2}. Such variability in
recruitment is a characteristic of many bivalve populations (Hughes,
1970), and at Skeffling, it appears to have been unrelated to the
numbers of breeding adults in the population.

The density of the 1973 year class at site 3 remained steady
throughout spring and summer 1974 but fell from nearly 20,000 m^{-2} to
5,000 m^{-2} in October, and to under 2,000 m^{-2} by April 1975. This
represents a 94% drop in numbers of this year class in this area
during their second year of life. During this period, numbers above
and below this area of the shore dropped similarly and this decrease
is, therefore, interpreted as mortality.

The depth distribution of <u>Macoma</u> in the sediment at Skeffling
was investigated three times during 1976 and the results are shown
in Figure 8. At all times of year, smaller animals were found in
the surface layers and larger animals progressively deeper. Animals
greater than 4.0mm long burrowed deeper in winter than in summer.
Smaller animals remained in the top 1cm of sediment throughout the
year. These findings and those of Reading and McGrorty (1978) support
the hypothesis of Vassallo (1971) that inhalent siphon length limits
burrowing depth.

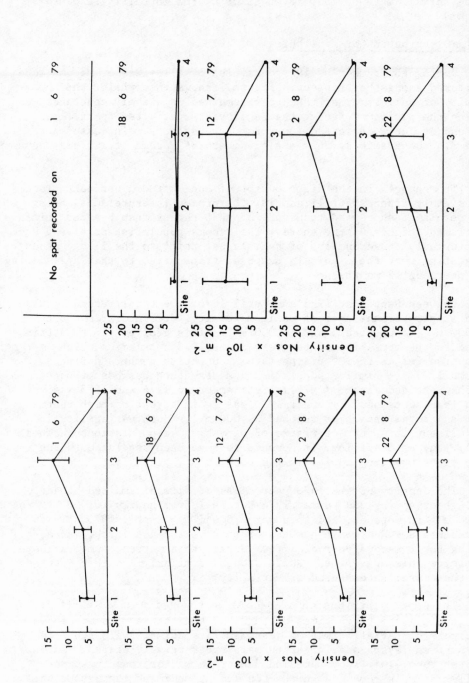

Figure 7. The total densities of 1+ and older Macoma balthica (left) and Macoma balthica spat (right) along the Skeffling transect with 95% confidence limits.

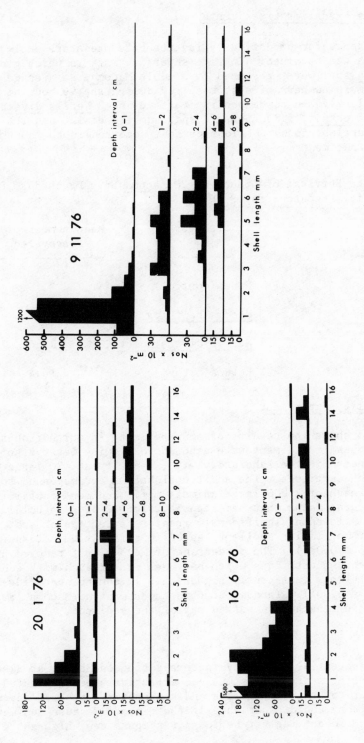

Figure 8. The numbers and sizes of <u>Macoma balthica</u> taken at different depths in the sediment at site 3 at Skeffling.

Invertebrate Predation Experiments

No evidence for predation by invertebrate predators on 1+ and older Macoma was obtained. In five separate tests in which groups of 20 Macoma spat were presented to 3 adult Macoma, an average of 9% reduction in numbers of spat resulted during the 96 hour test period (Table 2). In corresponding tests using 2 Nereis diversicolor and 2 Retusa obtusa, reductions in spat numbers of 12% and 18% respectively were obtained. No loss of spat occurred in the controls for these experiments.

Table 2.   The Survival of Groups of 20 Macoma Spat Exposed to Potential Predators for 96 Hours

Treatment No.	Predator	Mean number of spat recovered
1	None (control)	20
2	3 Macoma balthica	18.2
3	2 Nereis diversicolor	17.6
4	2 Retusa obtusa	16.4

Macoma Mobility Experiment

Table 3 shows the results of an experiment in which groups of five cartridges of sediment were exposed to flowing water after ten days maintenance in the laboratory with, and without, 20 large Macoma added to each cartridge. The additional Macoma corresponded to a density increase of 2,000 large animals m^{-2}. Sediment containing only the natural fauna that had been left for 10 days in holding conditions, allowed a significantly greater percentage of spat to be removed than similar sediment tested immediately on return from the field (p < 0.001). The percentage of Macoma spat removed by the water from those cartridges which had been pre-treated with additional Macoma were also significantly higher than in control cartridges of sediment (p < 0.001). Throughout this experiment no Macoma larger than 1.5 mm in length were washed out of the sediment.

DISCUSSION

Of the potential mortality factors acting on Macoma at Skeffling, the direct effects of temperature can effectively be eliminated. Macoma appear to be able to withstand temperatures several degrees higher than they could experience in the field. Because sediment temperatures did not lag far behind air temperature, high-shore

Table 3.   The Numbers and Percentages of <u>Macoma</u> Spat Removed From
            Cartridges of Sediment by One Hour Exposure to Flowing
            Water

	Number removed	Number remaining in sediment	Number removed as a percentage of total numbers	
C1 1	0	62	0	
2	0	44	0	Mean =
3	1	10	9	2.2%
4	0	50	0	
5	1	51	1.9	
C2 1	2	24	7.7	
2	6	15	28.6	Mean =
3	2	24	7.7	13.6%
4	5	22	18.5	
5	1	17	5.6	
Test				
1	16	9	64.0	
2	29	15	34.1	Mean =
3	28	14	66.7	52.8%
4	17	13	56.7	
5	14	19	42.4	

Control 1   no <u>Macoma</u> added to sediment, tested immediately on
             return to the laboratory.

Control 2   no <u>Macoma</u> added to sediment, tested after 10 days in
             holding conditions.

Test   20 large <u>Macoma</u> added to each cartridge, tested after
             10 days in holding conditions.

areas are likely only to experience longer periods of high temperature than low-shore areas, rather than higher temperatures. Kennedy
and Mihursky (1971) found that the upper lethal temperature for
<u>Macoma</u> lay between 31°C and 35°C depending on acclimation temperature.
These tests were, however, carried out over a period of 24 hours.
The present 6 hour tests correspond more closely to the length of
time during which <u>Macoma</u> in the intertidal zone might experience
high temperatures. de Wilde (1975) recorded poor growth and loss
of condition in <u>Macoma</u> kept at constant temperatures above 15°C and
heavy mortality at constant temperatures above 25°C. At Skeffling,
however, the highest water temperature recorded was 22°C and <u>Macoma</u>
could never have been exposed to higher temperatures for more than
about 10 hours.

It is doubtful whether appreciable mortality of Macoma at
Skeffling occurred as a direct result of low temperatures. Muus
(1967) recorded few mortalities of Macoma in Denmark at temperatures
similar to the lowest recorded at Skeffling. de Wilde (1975) found
that Macoma kept at 0°C maintained a high level of feeding activity.
From the data given by Crisp (1964) it appears that Macoma suffered
little, if any, mortality around British coasts directly due to the
low temperatures of the winter of 1962/3.

What may be more important than the direct effects of temperature are the interactions between temperature and other factors.
During winter, animals rendered moribund by very low temperatures
may not be able to remain deeply buried during windy weather and
may even become exposed during periods of high sediment mobility.
Such exposure would leave them more vulnerable to bird predation.

During periods when ice is present in water overlying the mudflats, scouring of the sediment surface by the ice, and consequent
movement of animals may take place. The migration of larger Macoma
towards the surface layers of the sediment during the summer is
probably a temperature-related phenomenon and would also render the
animals more vulnerable to capture by birds. Infection by the trema-
tode parasite Parvatrema affinis may also result in migration of
larger Macoma to the sediment surface (Swennen and Ching, 1974).
During the summer, however, the numbers of wading birds feeding on
the mud-flats is at a minimum but the Skeffling area is an important
feeding ground in the winter.

The depth to which Macoma burrow will influence their availability as prey for birds. At Skeffling, in June 1976, more than
99% of the Macoma population lay within the probing depth range of
Dunlin, which have a mean beak length of about 3cm which is the
shortest beak of any common avian predator of Macoma in the area.
Thus, other, larger species, could easily reach the entire Macoma
population in the summer. In January, however, 32% of Macoma lay
deeper than the maximum probing depth of Dunlin and 19% beyond the
probing range of Redshank, with a beak length of 4cm. At all times
of year, however, birds with longer beaks such as Curlew and Godwit,
would have access to the entire Macoma population. These conclusions
are the same as those reached by Reading and McGrorty (1978).

According to Reading and McGrorty (1978), 80% of the Macoma taken
by Knot in the Wash are in the size range 9 to 13mm. If this strong
preference for larger Macoma is shown by birds at Skeffling, then a
large proportion, often over 90% of the Macoma population will remain
unselected by birds, although accessible to them.

Whilst predation pressure from birds is only present when the
mud-flats are exposed by the tide and is probably much greater in
winter than summer, predation from shrimps, crabs and young flatfish

during periods of water cover will be greatest when these predators
are most abundant during the summer (Reise, 1977; Riley, 1979).
However, no assessment of the effects of their predation on the
Macoma population has yet been attempted.

Predation on Macoma spat by adult Macoma, Nereis and Retusa has
been demonstrated. Such predation could occur at all states of the
tide throughout the year. Once Macoma spat have entered their first
full summer's growth however, they will escape predation from these
invertebrate predators because they will be too large. Most Macoma
at Skeffling die during their first year of life and it is these
0+ Macoma that will be most vulnerable to predation by other mud-
dwelling invertebrates. Bird predation is probably greatest on
older animals. No quantitative evaluation of the effects of predation on the Macoma population is yet available. However, if predation is an important mortality factor, predation by invertebrates
or young flatfish on 0+ Macoma may be important, although de Vlas
(1979) found little evidence in the Wadden Sea where the fish fed
mostly on the siphon tips of Macoma.

Short-term changes in spat distribution occurred during the
two months immediately after spatfall when the young animals were
growing from a mean shell length of 300-400µm immediately after
settlement, to around 1mm in September at which size they overwintered.
Similar observations were made by Wolff and de Wolf (1977) which were
interpreted as the result of migratory behaviour. There is no evidence to suggest that a return movement down the shore takes place
and it is, therefore, suggested that the observed movement is passive
rather than a true migration. From the results of experiments with
flowing water, it is probable that 0+ animals remained vulnerable
to movement by water currents from the time of spatfall until the
time at which growth during the following spring allowed them to
avoid being moved. This would be either by virtue of being able to
burrow deeper, and so avoid the unstable surface layers of the
sediment, or by their increase in weight rendering them more resis-
tant to movement by water currents.

During the summer, the sediment surface was relatively stable
whereas erosion occurred during autumn and winter. Consequently,
the spat movement recorded during summer may have been quite small
compared with that which could occur during autumn and winter.
Because small Macoma must remain in close contact with the sediment
surface throughout the year, they are at all times exposed to movement with the mobile surface layers of the sediment whereas larger
animals may, under most conditions, be able to avoid such movement
by deeper burrowing. The appearance of dense bands of dead shells
of larger Macoma at the lower edge of the Spartina marsh after winter
storms however, suggests that under severe weather conditions the
entire Macoma population is vulnerable to movement.

The results of experiments with flowing water indicate that the feeding activities of larger Macoma make spat more vulnerable to being moved by water currents. This effect may be the result of a behavioural change of the spat when in the presence of larger Macoma, for example, they may move upwards in the sediment. Alternatively, it may be caused by spat being picked up by the inhalent siphons of adult Macoma and being ejected onto the sediment surface. A third possibility is that the working of the sediment by larger Macoma results in loosening of the sediment surface structure, making it more unstable when exposed to water currents. The large fluctuations in sediment height recorded in the region of maximum Macoma density support this latter argument.

If spat are moved by water currents, they may be transported to areas more or less favourable for survival than those from which they moved. At Skeffling, spat were transported to the site 3 area where sediment organic content and, therefore, food was more abundant than further down the shore. However, the possibility of food limitation cannot be ruled out because of the much higher densities of Macoma in this area than elsewhere on the shore. In being transported to an area of high adult density and, hence, greater sediment instability, spat may decrease their chances of survival if they are continuously being moved about on the surface of the sediment.

In summary, it would appear that the success of M.balthica in the lower Humber estuary can be attributed to its life-style and life-cycle being amenable to distribution by, and tolerance of, the physical features of the part of the estuary that the species inhabits.

REFERENCES

Anderson, S.S. 1972. The ecology of Morecambe Bay. II. Intertidal invertebrates and factors affecting their distribution. J. appl. Ecol. 2: 161-178.

Beukema, J.J., Cadee, G.C. and Jensen, J.J.M. 1977. Variability of growth rate of Macoma balthica (L.) in the Wadden Sea in relation to availability of food. In B.F. Keegan, P.O. Ceidigh and P.J.S. Boaden (eds), Proc. 11th Europ. mar. biol. Symp. 69-77. Pergamon Press.

Chambers, M.R. and Milne, H. 1975. The production of Macoma balthica (L.) in the Ythan estuary. Est. coast mar. Sci. 3: 443-455.

Crisp, D.J. 1964. The effects of the severe winter of 1962-63 on marine life in Britian. J. Anim. Ecol. 33: 165-210.

Eagle, R.A. 1975. Natural fluctuations in a soft bottom benthic community. J. mar. Biol. Ass. U.K. 55: 865-878.

Gilbert, M.A. 1975. Distribution, reproduction, feeding and growth of Macoma balthica in New England. Unpublished Ph.D. Thesis, University of Massachusetts.

Hughes, R.N. 1970. Population dynamics of the bivalve Scrobicularia

plana (da Costa) on an intertidal mud-flat in North Wales.
J. Anim. Ecol. 39: 333-356.

Kennedy, V.S. and Mihursky, J.A. 1971. Upper temperature tolerances
of some estuarine bivalves. Chesapeake Sci. 12: 193-204.

McLusky, D.S. and Allan, D.G. 1976. Aspects of the biology of Macoma
balthica (L.) from the estuarine Firth of Forth. J. moll. Stud.
42: 31-45.

Muus, B.J. 1967. The fauna of Danish estuaries and lagoons,
distribution and ecology of dominating species in the shallow
reaches of the mesohaline zone. Meddr. Danm. Fisk.-og
Havunders. 5: 1-316.

Muus, K. 1973. Settling, growth and mortality of young bivalves
in the Øresund. Ophelia 12: 79-116.

Ratcliffe, P.J. 1979. An ecological study of the intertidal
invertebrates of the Humber estuary. Unpublished Ph.D. Thesis,
University of Hull.

Ratcliffe, P.J. 1980. The distribution and mortality of the young
of Macoma balthica (L.) in the lower Humber estuary. Unpublished M.Sc. Thesis. Plymouth Polytechnic.

Reading, C.J. 1979. Changes in the downshore distribution of Macoma
balthica (L.) in relation to shell length. Est. coast. mar.
Sci. 8: 1-13.

Reading, C.J. and McGrorty, S. 1978. Seasonal variations in the
burying depth of Macoma balthica (L.) and its accessiblity to
wading birds. Est. coast mar. Sci. 6: 135-144.

Rhoads, D.C. and Young, D.K. 1970. The influence of deposit-feeding
organisms on sediment and community trophic structure. J. Mar.
Res. 28: 150-178.

Reise, K. 1977. Predator exclusion experiments in an intertidal
mud-flat. Helgoländer wiss. Meersunters. 30: 263-271.

Riley, J.D. 1973. The biology of young fish in the Humber estuary.
In N.V. Jones (ed.) University of Hull/Humber Advisory Group,
Joint Symposium. Limited publication, University of Hull.

Swennen, C. and Ching, H.L. 1974. Observations on the trematodes
Parvatrema affinis, causative agent of crawling tracks of
Macoma balthica. Neth. J. Sea Res. 8: 108-115.

Thorson, G. 1957. Bottom communities (sub-littoral or shallow
shelf). In J.W. Hedgepeth (ed.), Treatise on marine ecology
and paleoecology. Geol. Soc. Am. Mem. 67: 461-534.

Tunnicliffe, V. and Risk, M.J. 1977. Relationships between the
bivalve Macoma balthica and bacteria in intertidal sediments:
Minas Basin, Bay of Fundy, J. mar. Res. 35: 499-507.

Vassallo, M.T. 1971. The ecology of Macoma inconspicua (Broderup
and Sowerby, 1829) in central San Francisco Bay. Part II.
Stratification of the Macoma community within the substrate.
Veliger 13: 279-284.

Wilde, P.A.W.J. de 1975. Influence of temperature on behaviour,
energy metabolism, and growth of Macoma balthica (L.). In
H. Barnes (ed.) Proc. 9th Europe. mar. biol. Symp. 239-256.
Aberdeen University Press.

Wolff, W.J. and de Wolf, L. 1977. Biomass and production of zoo-
    benthos in the Grevelingen Estuary, The Netherlands.
    Est. coast. mar. Sci. 5: 1-24.

THE FEEDING AND SURVIVAL STRATEGIES OF ESTUARINE MOLLUSCS

D.S. McLusky[1] and M. Elliott[2]

[1]Department of Biology, University of Stirling,
Stirling, Scotland. FK9 4LA
[2]Forth River Purification Board, Estuary Laboratories,
Port Edgar, South Queensferry, Edinburgh

INTRODUCTION

The fauna of estuaries is characterised by having relatively
few species, the number decreasing within the estuary from both the
seaward and river end to reach a minimum of species at a salinity
of c. 5‰, whilst at the same time these few species may be extremely
abundant. The abundance of estuarine animals leads to the recogni-
tion of estuaries as one of the most productive natural habitats
(McLusky, 1971; in press). The high productivity of estuaries is
based on their ability to retain detritus material derived from the
sea, rivers or salt marshes. Feeding on this detritus are three
main taxa of animals, the annelids, the crustaceans and the molluscs.
The present review is concerned with the latter phylum, the molluscs,
two groups of which, the gastropods and the bivalves, form a most
conspicuous part of the estuarine macrofauna. As with other members
of the estuarine fauna there may be relatively few species of
molluscs in estuaries, but these species may be very abundant indeed.

In the present paper we intend to review some of the features
of molluscs which have led to their abundance in estuaries, and
some of the problems of estuarine life which have prevented other
mollusc species from entering these environments. In order to
analyse more carefully which features of the estuarine ecosystem
control the growth and success of mollusc populations, we will also
examine the variation in the success of five mollusc species (Macoma
balthica (L.), Cerastoderma (=Cardium) edule (L.), Mya arenaria (L.),
Retusa obtusa (Montagu) and Hydrobia ulvae (Pennant) studied in
the Forth estuary. Each of these species occupies a different
ecological niche within the estuarine ecosystem and we will hope
to show how the strategies of the different species are adapted

in response to the environmental factors within the estuary.

GENERAL REVIEW

The importance of the molluscs to the functioning of estuarine ecosystems has been shown by many authors. Within the Grevelingen estuary, Wolff and de Wolf (1977) showed that of the total production of 57.43 g m^{-2} yr^{-1} due to the macrofauna, 53.69 g m^{-2} yr^{-1} (93%) could be attributed to five mollusc species (Littorina, Hydrobia, Cardium, Macoma and Mytilus). Beukema (1979) showed that the mean biomass of the macrozoobenthos in the Balgzand part of the Dutch Wadden Sea between 1971 and 1978 was 19.6 g m^{-2} yr^{-1}, with 12.3 g m^{-2} yr^{-1} attributed to the molluscs Mya, Cardium, Macoma and Mytilus. Taking the Wadden Sea as a whole, Beukema (1976) showed that 66% of the total macrofaunal biomass was due to molluscs (65% bivalves, 1% gastropods). Within the Lynher estuary, Warwick, Joint and Radford (1979) found a smaller proportion of the macrofaunal biomass and production due to molluscs, with 26% of production being due to Mya, Scrobicularia, Cardium and Macoma. Burke and Mann (1974) showed that molluscs were the chief primary consumers within the Petpeswick Inlet estuary of Nova Scotia, and Dame (1976) found that the inter-tidal oyster (Crassostrea virginica) was the main primary consumer within the North Inlet estuarine ecosystem of South Carolina. Hibbert (1976) emphasised the contribution of bivalve molluscs to other trophic levels in his study of 60 hectares at Hamble Spit, where he calculated that predators removed 10.7 tonnes yr^{-1} of bivalves, leaving 18.2 tonnes yr^{-1} to pass to the scavenger-decomposer food chain.

The distribution of mollusc species within estuaries has been discussed by several authors (Green,1968; Muus,1967; McLusky,1971) who have reported a gradation of species within estuaries with a maximum number of species at the marine end and a progressive reduction, moving into the estuary, as each species approaches a minimum salinity at which it can survive. From the records of Wolff (1973) one can distinguish four groups of estuarine molluscs. Firstly, there are the marine species which are confined to salinities greater than 25‰, and which have pelagic larvae (e.g.: Buccinum undatum, Venerupis pullustra, Mactra corallina, Abra alba, Abra tenuis, Angulus tenuis). Secondly, the euryhaline species which can tolerate salinities down to 10-18‰, and which have pelagic larvae (e.g.: Cardium edule, Mytilus edulis). Thirdly, the estuarine species which can tolerate conditions down to 5-10‰ salinity and which may have non-pelagic larval stages (e.g.: Littorina saxatilis) or pelagic larval stages (e.g.: Macoma balthica, Mya arenaria). Fourthly, there are the freshwater species present at the river end of the estuary, which may live at salinities of up to 5‰ (e.g.: Potamopyrgus jenkinsi). The presence of non-pelagic or weakly swimming larvae in estuarine species as well as the ability to utilise currents is generally considered as an adaptation to avoid the larvae being carried out of the estuary (Wood & Hargis, 1971; Scheltema, 1976).

Robertson (1964) has reviewed the osmotic and ionic regulation abilities of the molluscs and has shown that marine molluscs are invariably isosmotic with sea water although they do show ionic regulation. The majority of euryhaline molluscs penetrate into estuaries from the seaward end and are isosmotic, allowing the osmotic concentration of their body fluids to decrease as the salinity of the medium decreases. Within the gastropods, Littorina littorea does show some hyperosmotic regulation in 8-16‰, and Hydrobia ulvae and Potamopyrgus jenkinsi also show hyperosmotic regulation below 16‰. When exposed to freshwater P.jenkinsi is able to continue an active life, but H.ulvae is forced to withdraw into its shell (Todd, 1964).

Within the bivalves, the normal situation appears to be for the body fluids to be isosmotic with the medium in all salinities until a lethal lower limit is reached. Most bivalves can, however, withstand short exposure to low salinities by shutting their shell and trapping high salinity water inside. The freshwater mussels, Anodonta and Dreissena, which inhabit the low salinity waters of the inner Baltic Sea are isosmotic with the medium at the very low salinity of 2.3‰. Above that they are isosmotic and below are slightly hyperosmotic. Mytilus edulis, Cardium edule, Macoma balthica and Mya arenaria penetrate the Baltic Sea reaching salinities of 2-5‰, which are lower salinities than they tolerate in the more variable salinities of estuaries. All these species, however, show a marked reduction in growth rate and maximum size at these low salinities and it must be suggested that these conditions represent the fringe of the species' distribution. Although bivalves do not show osmotic regulation of the body fluids in low salinities, they do show anisosmotic intra-cellular regulation (Gilles, 1979), which provides a degree of protection for the cells when the animals are exposed to low salinities.

Apart from salinity, the other dominant environmental feature of estuaries is the presence of mud which as a substratum may be difficult to live in and is also the main factor in maintaining turbid water conditions. But whilst mud may be a difficult physical medium for an animal to adapt to, it is generally a rich biological habitat with an abundance of micro-organisms and a much higher total organic content than other substrata. Mud may tend to clog the feeding and respiratory mechanisms, especially of molluscs and, indeed, the absence of aspidobranch gastropods from muddy areas has been attributed to the problems of clogging respiratory mechanisms (Yonge, 1953). As we have seen above, the commonest molluscs within estuaries are the bivalves, which feed with the aid of highly developed gills.

Whether the bivalve mollusc is deposit-feeding on a muddy substratum or filter-feeding on the particles contained in muddy estuarine waters, there remains the problem of separating the fine particulate nutritional material from the fine particulate inorganic fragments of silt and clay. The main adaptation of bivalves to such a problem is the development of siphonal tubes which serve to limit the inflow of water to the gills and also to act as some means of pre-selection. Yonge (1953) has remarked on a distinction between the siphons of suspension-feeding bivalves and the siphons of deposit feeding bivalves. In the case of the former, there is a tendency towards the development of fused siphons, with the gradual atrophy of the foot whilst the siphons grow longer. The suspension feeding bivalves tend to bury deeper with increasing age, and as the siphons become covered in a strong periostracum they can no longer be retracted into the shell. Examples of such a mode of life are provided by Mya arenaria and M.truncata, as well as Lutraria lutraria and Schizothaerus nuttallii.

By contrast, the deposit-feeding bivalves develop long and separate siphons which can reach out from the bivalve far over the surrounding surface to pick up food items. The siphons remain active and extensile, and the foot is generally large and active. Within the animal, the labial palps are large and have special mechanisms for rejection of waste. Yonge (1949) has shown, within the Tellinacea, that as the sediment particle size decreases there is an increase in the size of labial palps in relation to gill size. Mantle folds separate incoming and outgoing material and the stomach is also specialized to deal with great quantities of bottom material. Examples of such molluscs are Macoma balthica or Scrobicularia plana.

The third mode of feeding used by estuarine molluscs is the grazing method adopted by the gastropods, such as Littorina and Hydrobia species, which use their radula to remove the algal film from rocky outcrops, pebbles or finer particles.

The distinction between these three feeding mechanisms may, however, not be quite as clear as Yonge suggested. For example, de Wilde (1975) has shown that Macoma balthica can spend 10-40% of its life suspension-feeding and 60-90% of its life deposit-feeding. Tunnicliffe & Risk (1977) have shown that although Macoma feeds on bacteria within the sediment, it must supplement its diet by suspension-feeding during high tide in order to acquire sufficient protein. Hughes (1969) has described the adaptations of Scrobicularia plana to deposit-feeding, including gill modifications which reduce the risk of choking and increase the efficiency of transfer of material to the palps, however, he also found that the animal spends part of its life suspension-feeding and could retain suspended food particles down to 4μm in size. Rasmussen (1973) also noted the flexibility in the mode of feeding of Macoma balthica between deposit and suspension feeding, and he further suggested that Mya arenaria is a suspension feeder when covered by a considerable depth of water, but as the water level drops with the tide, it can switch to being a deposit feeder by drawing in food particles from the surrounding mud surface. This flexibility in the mode of feeding as a characteristic of estuarine molluscs was also shown by Fenchel, Kofoed and Lappalainen (1975) who showed that Hydrobia ulvae can ingest large particles, browse on surfaces, and also, probably utilise mucus for trapping micro-organisms, and can thus embrace deposit, grazing and suspension feeding.

The importance of the tolerance of muddy conditions as well as low salinities for estuarine molluscs can be clearly seen in a comparison of the oysters Ostrea edulis and Crassostrea virginica (Yonge, 1960). Both Ostrea and Crassostrea are euryhaline, but Crassostrea can tolerate salinities down to 12‰, whilst Ostrea can only tolerate salinities down to 23‰. Ostrea cannot tolerate conditions as turbid as those accepted by Crassostrea. Crassostrea is able to tolerate turbid conditions with the assistance of well-developed mantle margins which mimic siphonal tubes, with more discriminatory palps, with additional cleaning currents, and with more quick muscle fibres in the adductor muscle. In this comparison of oysters can be seen the many adaptations required for successful estuarine life.

The pollution ecology of estuarine molluscs has been reviewed recently by Menzel (1979), and is not considered in this paper.

STUDIES OF FORTH MOLLUSCS

The population and production ecology of five species of molluscs has been studied at fourteen stations on Torry Bay, a large inter-tidal area in the estuarine Firth of Forth, eastern Scotland, over the period January 1975 to February 1977. The species are the pre-dominantly deposit-feeding bivalve Macoma balthica the predominantly suspension-feeding Mya arenaria and Cardium (=Cerastoderma) edule, and the gastropods Hydrobia ulvae (detritus feeder) and Retusa obtusa (predatory on Hydrobia and spat of bivalves).

The fourteen stations covered a range of tidal heights and substratum types within Torry Bay (Fig. 1). The Torry Bay area experiences a salinity range of 35.4 - 22.3‰, and is situated on the northern shore of the Forth some considerable distance from major polluting sources (McLusky, Elliott & Warnes, 1978). The full study, described in Elliott (1979), includes an evaluation of the productivity of each of the mollusc species, and a detailed comparison between the performance of each species at each station by a series of multiple regression and other analyses. This comparison permits a specification to be made of the preferred sites and thus an elucidation of which strategies are most success-

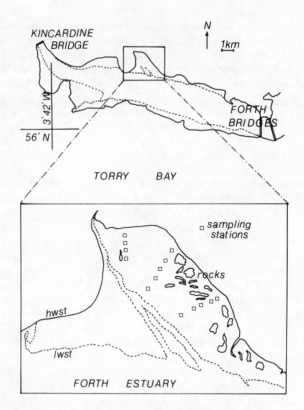

Figure 1. Map showing location of Torry Bay on north shore of Forth
estuary, Scotland, with insert to show position of sampling stations
on the intertidal area.

ful in these populations. In the present paper we will describe
which conditions were found to be most suitable for each species,
and which ecological strategies were most effective.

Macoma balthica

     The population of Macoma balthica within Torry Bay was normally
between 100 - 400 (> 2.5 mm) animals m^{-2}, but with up to 2000 m^{-2} in
preferred areas. The largest populations occurred in areas of high
levels of particulate carbohydrate and with large spat populations,
at 3.0 - 3.5 m above chart datum, an immersion period of 5.7 to
7.5 h at neap tides and 6.0 to 6.7 h at spring tides, and in the finer
sediments. These findings are in agreement with Tunnicliffe & Risk
(1977), Warwick and Price (1975) and Wolff and de Wolf (1977). In
addition, the population was optimal in sediments with 2.5%C, 0.12%N
and C/N ratio values of 20 - 23, suggesting that Macoma showed a
preference for relatively un-nutritious muds, i.e. those muds which
contained a high proportion of undegraded carbon as well as a relatively high percentage of nitrogen.

The importance of a fine substrate in providing food for
M.balthica was demonstrated by Newell (1965), Ankar (1977) and
Tunnicliffe and Risk (1977). As has been found in the present study,
these authors found a correlation between the density of Macoma and
the proportion of silt and clay in the sediments; however, they
found no overall correlation between the density and sediment particle
size. These authors assumed that the bacterial content of a sediment
is directly related to the nitrogen and silt and clay content, and
that the carbon content is a direct measure of undecomposed detritus
which acts as a substrate for the microfauna. It is likely that the
bacteria were initially attracted by the high proportion of silt and
clay. Once established, the bacterial flora may be enhanced by the
supply of bivalve faecal pellets (i.e. a "gardening strategy"). The
importance of fine sediments for Macoma is also emphasised by
Beukema et al., (1977), who found that the primary production of
microphytobenthos was greatest in fine sediments, and that this
microphytobenthos was the main food of Macoma in the Wadden Sea area.
So we may suggest that the strategy of selecting finer substrata by
Macoma is of value to this predominantly deposit-feeder because
such substrata support a richer microflora.

The nature of the substratum, however, is not the only factor
determining Macoma density. The present study has shown that the
quantity of particulate carbohydrate in the water influenced the
population size. It is not possible to indicate a direct cause and
effect relationship in this study, but in common with other works
(Tunnicliffe and Risk, 1977) and from our own experimental observations, we may suggest that Macoma can supplement its diet by suspension-feeding. Macoma balthica produces planktonic spat which have
been found in the present study to settle predominantly on fine
sediments. It is possible that the larvae were either attracted to
the bacterial film on fine sediments, or that gregarious settlement
has occurred, or that hydrographic concentration (Turner, 1953)
occurred with maximal settlement in areas with low current speed
and, therefore, fine sediments.

The large fluctuations in M.balthica abundance within the study
area may be usual in an estuarine situation. Myren and Pella (1977)
suggest that estuaries are characterised by simple food chains, and
the presence of a few species, each with a short life span and tolerant of wide physiological changes. They argue that because the
estuarine ecosystem is held in immaturity by harsh environmental
changes the species present have large temporal variations in abundance. Macoma, however, has features which suggest greater population
stability; it may either suspension or deposit feed, thus, ensuring
a more continuous food availability; it is comparatively long lived
(10 years here) and is tolerant to relatively large scale changes in
the physical environment. The large fluctuations in abundance need
not, therefore, be attributed to harsh environmental changes, but may
be due to variability in the success of its mode of reproduction and
to differences in food availability and physical characteristics

within the intertidal area.

Cardium edule and Mya arenaria

These predominantly suspension-feeding bivalves showed several similarities in their ecology and may be considered together. Adult Cardium (maximum population size = 3000 m^{-2}, 200 m^{-2} > 2 mm) were found mainly in high-shore areas of steeply sloping coarse sediment with large amounts of carbohydrate in the water. By contrast, the spat of Cardium settled predominantly in areas of fine mud above the mid to low-shore level, with a shallow slope, high amounts of silt and clay in the sediment and high interstitial salinity. It is, therefore, necessary to consider the adults and spat separately because of their different ecological preferences (Savilov, 1953). It would appear that a low-shore fine sediment area is favourable for passive settlement of the poorly swimming larvae, due to weaker water currents in such areas. These areas, however, may be deleterious to spat after settlement as they are associated with a large silt inflow, which would lead to the smothering of the spat together with a lower supply of suspended phytoplankton food, and a slower removal of waste material (Pratt, 1953). To avoid these problems secondary movement of spat may occur after settlement.

Adult Mya had the greatest densities around the mid-tide level on fine sediments with a shallow slope and high water organic matter levels. The areas of greatest biomass were at the mid to upper-shore levels. The distribution of adult Mya was approximately the same as the distribution of Mya spat, suggesting minimal transport. The Mya population was, therefore, found to favour muddy areas where the substratum was stable, which is in agreement with Wolff (1973). The population density of Mya at Torry Bay had a mean of 220 m^{-2} (30 - 40 m^{-2} > 2 mm) which is regarded by Jones (1960) as constituting a well-populated area.

The growth of Cardium at Torry Bay varied spatially and was dependent on environmental conditions. The growth rate of young Cardium was greatest at the low-shore muddy areas where few large cockles were present, but the growth rate later in life was greatest at the upper tidal areas with coarser sediments. However, the size of older animals decreased with increasing tidal height and they achieved the greatest size at mid tidal levels which did not support the greatest biomass of adult Cardium. Other workers (Stephen, 1930; Verwey, 1952; Barnes, 1972) have stressed the importance of submersion time as controlling the feeding and growth of suspension feeders. But the present study suggests that tidal height should be considered together with sediment type and its related characteristics, as well as the associated water movements.

The deleterious effect of silt on suspension-feeding, whereby the rate of filtration and feeding are decreased and pseudofaeces

production and energy expenditure increased, has been documented by
Johannessen (1973). The present study has shown passive settlement
of Cardium and Mya spat in muddy "quiet" areas, but in the case of
Cardium greater subsequent growth has occurred in non-muddy areas
which have increased water currents which increase food availability
and oxygen exchange. The adults of Mya were found in similar areas
to those where spat settled which may reflect the lesser mobility of
Mya compared to Cardium, but may also be related to the ability of
Mya to deposit-feed at low tide (Rasmussen, 1973)), which Cardium
is unable to do. It might, therefore, be suggested that adult
Cardium are committed to suspension-feeding, but that the feeding
strategy of adult Mya is more variable and that it is prepared to
utilise deposit as well as suspension feeding.

Hydrobia ulvae

The Hydrobia ulvae population at Torry Bay had its centre
(40,000 individuals m^{-2}) at an upper shore level of 4-5 m above
chart datum, in sediments which suggest that the species had a
preference for greater wave action, and in an area which had a
supply of detritus as food provided by beds of Zostera marina and
associated epiphytes and microflora. Smaller specimens of Hydrobia
were found at lower tidal levels, which may be due to one or more
of several reasons: firstly, passive settlement of larvae would
occur in areas where water currents were slowest; secondly, the
young may be avoiding direct competition with the adults; or
thirdly; the young may have different food, exposure, salinity or
substratum preferences from the adults.

The evidence for the mobility of Hydrobia larvae is contradictory. Fish and Fish (1976) report a planktotrophic larval stage
which is pelagic for about 4 weeks, whereas Anderson (1971) quotes
suggestions that there is little or no free-swimming activity. The
present study, with a different distribution pattern for young compared to adults, does suggest that some mobility occurs, but is
unable to discern whether the settlement of young is passive, or
whether an active choice of settlement site is involved.

The distribution of the adult Hydrobia ulvae at Torry Bay was
primarily influenced by tidal height and the presence of Z.marina.
These findings are not wholly compatible with Newell (1965) who
found a high correlation between the particle size, amount of
organic matter and numbers of H.ulvae. It may be valid to suggest
that in fine mud and bare sand H.ulvae will have to ingest particulate matter to utilise the microflora on the particles' surface.
This method of feeding has a low assimilation efficiency (Fenchel,
1972) and consequently requires high energy expenditure. It may
be advantageous, therefore, for H.ulvae to adopt the feeding strategy
of browsing on microfloral epiphytes from the surface of Z.marina
and its detritus where possible. The fact that the H.ulvae popula-

tion at Torry Bay showed few relationships with the sediment
particle size suggests these animals should be regarded primarily
as a herbivore or grazer, rather than as a selective (Fenchel, 1972)
or indiscriminate (Newell, 1965) deposit feeder. It is, therefore,
postulated that when a more nutritious food source is available
near its optimal tidal range, that H.ulvae will adopt the feeding
strategy of grazing as a herbivore or selective detrital-feeding
rather than to deposit-feed indiscriminately.

Retusa obtusa

Preying upon the population of Hydrobia ulvae at Torry Bay is
a population of the gastropod Retusa obtusa which occurred at densities of approximately 250 m^{-2}. Retusa of size range 3.7 - 4.8 mm
were found to prey upon Hydrobia of size 1.4 - 2.3 mm. The life
cycle of Retusa in the Forth estuary is similar to that noted by
Smith (1967).

The predator/prey relationship provides a means of assessing
the feeding strategy of Retusa. Ecological efficiency, as the ratio
of the food intakes of each successive trophic level is usually of
the order of 10-20% (Steele, 1974). In the present case the ecological efficiency of Retusa preying on Hydrobia was between 0.93 and
1.67%, which is an order of magnitude lower than for systems where
the prey species is almost totally supporting the predator population. Retusa has a direct development (Thompson, 1976) without any
pelagic phase. On the Forth estuary, the animal has only been found
in a small number of discrete populations and although Hydrobia has
an extensive distribution on the estuarine Firth of Forth, this is
not reflected by the distribution of Retusa. In comparing the reproductive strategy of these two gastropods it would seem that the pelagic development of Hydrobia provides a more effective means of
colonising the estuary than the non-pelagic development of Retusa,
which may be contrasted with statements (e.g. Green, 1968) that the
suppression of the free-swimming larval stage may be of value to
estuarine animals.

CONCLUSION

The survival strategies for estuarine molluscs are, thus, a
tolerance of low salinities, a tendency towards non-pelagic larvae,
and an adoption of a flexible mode of feeding designed to cope with
the problems of a muddy environment.

Earlier work (loc. cit.) have classified the feeding strategies
of estuarine molluscs into three separate categories; deposit,
suspension and rasping feeding. The study of the five main mollusc
species of the Forth estuary, and an increasing amount of material
in the literature, points to the ability of estuarine molluscs to
utilise a number of different food sources, either at different

times in their life history, or at different phases of the tide.
It must be emphasised that there may be variability both between
populations of a species within a single estuary, and between
populations in different estuaries. Estuarine molluscs would seem
to be characterised by the adaptability of their survival and feeding strategies. In conclusion, we propose that it is this flexibility of feeding strategies which is the hallmark of a successful
estuarine mollusc.

REFERENCES

Anderson, A. 1971. Intertidal activity, breeding and the floating
    habit of Hydrobia ulvae in the Ythan estuary. J. mar. biol.
    Assn. U.K. 51: 423-437.
Ankar, S. 1977. The soft bottom ecosystem of the northern Baltic
    proper with special reference to the macrofauna. Contr. Askö
    Lab., 19: University of Stockholm. 1-62.
Barnes, R.S.K. 1972. Commercial species in the Solent area. II
    The abundance and population structure of cockles (Cerastoderma
    edule and C.glaucum) in Southampton Water 1970-72. C.E.R.L.
    Rep. No. RD/L/R 1786. (Mimeo publication).
Beukema, J.J. 1976. Biomass and species richness of the macro-
    benthic animals living on the tidal flats of the Dutch Wadden
    Sea. Neth. J. Sea Res., 10: 236-261.
Beukema, J.J. 1979. Biomass and species richness of the macrobenthic
    animals living on a tidal flat area in the Dutch Wadden Sea:
    Effects of a severe winter. Neth. J. Sea Res., 13: 203-223.
Beukema, J.J., G.C. Cadee and J.J.M. Jansen, 1977. Variability of
    growth rate of Macoma balthica (L.) in the Wadden Sea in
    relation to availability of food. In B.F. Keegan et al.,
    (eds). Biology of benthic organisms, Pergamon Press, Oxford,
    69-77.
Burke, M.V. and K.H. Mann, 1974. Productivity and production:
    biomass ratios of bivalve and gastropod populations in an
    eastern Candadian estuary. J. Fish. Res. Bd. Can., 31: 167-177.
Dame, R.F. 1976. Energy flow in an intertidal oyster population.
    Est. and Coastal Mar. Sci., 4: 243-253.
de Wilde, P.A.W.J. 1975. Influence of temperature on behaviour,
    energy metabolism and growth of Macoma balthica. In H. Barnes
    (ed.) Proc. 9th Europ. Mar. Biol. Symp., Aberdeen Univ. Press,
    239-256.
Elliott, M. 1979. Studies on the production ecology of several
    mollusc species in the estuarine Firth of Forth. Ph.D. Thesis,
    University of Stirling.
Fenchel, T. 1972. Aspects of decomposer food chains in marine
    benthos. Verh. dt. zool. Ges. (65 Jahresvers.), 14-23.
Fenchel, T., L.H. Kofoed and A. Lappalainen, 1975. Particle size
    selection of two deposit feeders - Corophium volutator and
    Hydrobia ulvae. Mar. Biol., 30, 119-128.
Fish, J.D. and S. Fish, 1977. The veliger larva of H.ulvae with

observations on the veliger of L.littorea (Mollusca: Prosobranchia). J. Zool. Lond., 182: 495-503.

Gilles, R. 1979. Mechanisms of osmoregulation in animals: maintenance of cell volume. J. Wiley, New York. 667 pp.

Green, J. 1968. The biology of estuarine animals. Sidgwick & Jackson, London. 401 pp.

Hibbert, C.J. 1976. Biomass and production of a bivalve community on an intertidal mudflat. J. exp. Mar. Biol. Ecol., 25: 249-261.

Hughes, R.N. 1969. A study of feeding in Scrobicularia plana. J. mar. biol. Assn. U.K., 49: 805-823.

Johannessen, O.H. 1973. Population structure and growth of Venerupis pallustra (Montagu). Sarsia, 52: 97-116.

Jones, B.W. 1960. On the biology of Mya arenaria. Rep. Challenger Soc., 3: No. 1223.

McLusky, D.S. 1971. The ecology of estuaries. Heinemann Educational Books, London. 144 pp.

McLusky, D.S. in press. The estuarine ecosystem. Blackies, Glasgow. 160 pp.

McLusky, D.S., M. Elliott and J. Warnes 1978. The impact of pollution on the intertidal fauna of the estuarine Firth of Forth. In McLusky, D.S. and A.J. Berry (eds), Physiology and Behaviour of Marine Organisms, Pergamon Press, Oxford. 203-210.

Menzel, W. 1979. Clams and Snails. In Hart, C.W. and S.L.H. Fuller (eds), Pollution Ecology of Estuarine Invertebrates, Academic Press, New York, 371-390.

Muus, B.J. 1967. The fauna of Danish estuaries and lagoons. Distribution and ecology of dominating species in the shallow reaches of the mesohaline zone. Meddr. Danm. Fisk. Havunds. Ny Serie, 5: (1), 316 pp.

Myren, R.T. and J.J. Pella, 1977. Natural variability in the distribution of an intertidal population of Macoma balthica subject to potential oil pollution at Port Valdez, Alaska. Mar. Biol. 41: 371-382.

Newell, R. 1965. The role of detritus in the nutrition of two marine deposit feeders, Hydrobia ulvae and Macoma balthica. Proc. Zool. Soc. Lond., 144: 25-45.

Pratt, D.M. 1953. Abundance and growth of Venus mercenaria and Callocardia morrhuana in relation to character of bottom sediments. J. mar. Res., 12: 60-74.

Rasmussen, E. 1973. Systematics and ecology of the Isefjord marine fauna (Denmark). Ophelia, 11: 1-507.

Robertson, J.D. 1964. Osmotic and Ionic Regulation. In Wilbur, K.M. and C.M. Yonge (eds), Physiology of Mollusca, Academic Press, London. 283-308.

Savilov, A.I. 1953. The growth and variation in growth of the White Sea invertebrates; M.edulis, M.arenaria and B.balanoides. Tr. Inst. Oceanology Acad. Sci. U.S.S.R. 7: 198-258.

Scheltema, R.S. 1976. Relationship of larval dispersal, gene flow and natural selection to geographic variation of benthic

invertebrates in estuarine and coastal regions. In L. Cronin
(ed), Estuarine Research, 1: Academic Press, New York, 372-391.
Smith, S.T. 1967. The evolution and life history of Retusa obtusa
(Montagu) (Gastropoda, Opisthobranchia). Can. J. Zool., 45:
397-405., 737-764.
Steele, J.H. 1974. The structure of marine ecosystems. Harvard
Univ. Press, Cambridge, Mass. 128 pp.
Stephen, A.C. 1930. Studies of the Scottish Marine Fauna:
Additional observations on the fauna of the sandy and muddy
areas of the tidal zone. Trans. R. Soc. Edin., 56: 521-535.
Thompson, T.E. 1976. Biology of Opisthobranch Molluscs. Ray Society,
London.
Todd, M.E. 1964. Osmotic balance in Hydrobia ulvae and Potamopyrgus
jenkinsi (Gastropoda, Hydrobiidae). J. Exp. Biol., 41: 665-677.
Tunnicliffe, V. and M.J. Risk, 1977. Relationships between the
bivalve Macoma balthica and bacteria in intertidal sediments.
J. mar. Res., 35: 499-507.
Turner, H.J. 1953. A review of the biology of some commercial molluscs of the east coast of North America. 6th Rep. Invest. Shell-
fish. Mass. Div. Mar. Fish. Dep. Nat. Res. Comm. Mass., Boston.
Verwey, J. 1952. On the ecology of the distribution of the cockle
and mussel in the Dutch Wadden Sea, their role in sedimentation
and the source of their food supply. Arch. Neerl. de Zool.,
10: 171-239.
Warwick, R.M., I.R. Joint and P.J. Radford, 1979. Secondary
production of the benthos in an estuarine environment. In:
R.L. Jefferies and A.J. Davy (eds), Ecological Processes in
Coastal Environments, Blackwell Sci. Pub., Oxford. 429-450.
Warwick, R.M. and R. Price 1975. Macrofauna production in an
estuarine mud-flat. J. mar. biol. Assn. U.K., 55: 1-18.
Wolff, W.J. 1973. The estuary as a habitat. An analysis of data
on the soft-bottom macrofauna of the estuarine area of the
rivers Rhine, Meuse and Scheldt. Zoologische Verhandelingen,
126: 1-242.
Wolff, W.J. and L. de Wolf 1977. Biomass and production of zoo-
benthos in the Grevelingen estuary, The Netherlands. Est.
Coastal Mar. Sci., 5: 1-24.
Wood, L. and W.J. Hargis, Jr. 1971. Transport of bivalve larvae in
a tidal estuary. In: D.J. Crisp (ed), 4th Europ. Mar. Biol.
Symp., Cambridge Univ. Press. 29-44.
Yonge, C.M. 1949. On the structure and adaptations of the Tellinacea,
deposit-feeding Eulamellibranchia. Phil. Trans. R. Soc., 234B:
26-76.
Yonge, C.M. 1953. Aspects of life in muddy shores. In: R. Elmhirst
(ed). Essays on Marine Biology, Oliver & Boyd, Edinburgh, 29-49.
Yonge, C.M. 1960. Oysters. Collins, London. 209 pp.

THE SURVIVAL, BEHAVIOUR AND RESPIRATORY PHYSIOLOGY OF CRANGON

VULGARIS (FABR.) IN THE POLLUTED THAMES ESTUARY

R.W. Sedgwick

Severn-Trent Water Authority
Regional Laboratory
Meadow Lane Nottingham   NG2 3HN

INTRODUCTION

In the Thames estuary the autumn migration of Crangon vulgaris
is unusual in bringing the shrimp in a landward rather than a seaward direction. Approaching the middle reaches, following along a
positive temperature gradient the population is confronted by an
increasingly more hostile environment as salinity and dissolved
oxygen (D.O.) levels decline near the nadir of the D.O. sag. The
ability of the shrimp to penetrate and successfully populate the
polluted estuary depends upon its powers of resistance to simultaneous osmotic stress and progressive hypoxia, particularly under
unfavourably high temperatures. During the period 1967-1973 the
regular migrations of C.vulgaris into the Thames estuary were
monitored quantitatively at West Thurrock (Huddart, 1971; Sedgwick,
1978) and subsequently a comprehensive laboratory study was made of
the shrimp's potential for continued survival in the multiple stress
environment in an attempt to predict shrimp distributions from water
quality data. In this paper attention is focussed on the conditions
which existed during and immediately following the migration, when
temperatures were still high, though falling, shrimp were in abundance, and the most extensive fluctuations in salinity and D.O. were
expected.

The study involved investigations into three main areas:

1.   An assessment of potential short and medium term survival
      capability.

2.   Examination of the potential energy available for dissipation
      in external activity as indicated by oxygen consumption.

123

3.   Assessment of the behavioural response of the shrimp to
      directionally applied D.O. depletion.

METHODS

General

     The shrimps were obtained from the Thames estuary at Tilbury and
were selected for size (5.0-6.0 cm total length) and sex. Only
females in intermoult were used. All individuals were allowed to
acclimate to constant conditions for 7 days before testing. They
were fed live Tubifex costatus during acclimation but not during the
experiments.

     Static water bioassays were carried out within enclosed glass
tanks of dimentions 30 cm x 20 cm x 20 cm, each incorporating an
internal sand and charcoal substrate filter. The tanks accommodated
12 individuals and usually four replications were made at each level
tested. Partial pressures of D.O. were controlled and maintained
by diffusing appropriate O_2/N_2 mixtures (British Oxygen Ltd), and
temperatures held at 20°C ± 0.2°C by immersion in water baths.
Salinity tolerance tests were continued for at least 72 hours but
when gas mixtures were employed practical considerations dictated
a maximum duration of 48 hours.

Salinity and Hypoxic Tolerance

     Salinity and hypoxic tolerance was determined using standard
graphic log-probit methods to obtain LC50 values (Litchfield, 1949;
Litchfield and Wilcoxon, 1949). In multiple stress experiments the
median survival or resistance time was adopted as the response
criterion. Analysis of these data was undertaken by fitting a
second degree polynomial and generating the response surface (Box,
1956).

Oxygen Consumption

     The oxygen consumption of individual shrimps was monitored at
20°C in a continuous flow, recirculating respirometer adapted from
Birtwell (1972). The respirometer cell was a 2.5 cm diameter perspex
cylinder containing washed sand into which the shrimp was encouraged
to bury itself by exposure to bright illumination during the experiments. Input D.O. levels were determined by dispersing O_2/N_2 gas
mixtures into the reservoir and output D.O. measured by a D.O.
electrode. At least one hour was allowed for the shrimp to become
adjusted to any D.O. change before recordings were taken. The control
of activity was imperative for meaningful interpretation of the data
and three states were recognised:

1.   Standard oxygen consumption was the minimum D.O. uptake over
     a test period when shrimp were quiescent and buried in the
     sand.

2.   Routine oxygen consumption was the maximum D.O. uptake of
     unstimulated shrimp over a test period.

3.   Active oxygen consumption was the maximum D.O. uptake obtained
     when shrimp were stimulated into continuous 'digging'
     behaviour by repeated manual rotation of the cell around its
     long axis, back and forth through 90°. This process caused
     the sand to be in motion beneath the animal and encouraged
     vigorous pleopod action which continued as long as the
     rotation was maintained.

     Shrimp chosen for oxygen consumption experiments were strictly
selected for size at 1.4 g wet weight. The relationships between
oxygen consumption and ambient D.O. were based on the response of
at least 10 shrimp in each case.

Assessment of Avoidance Behaviour

     The reactions of shrimp to rapid D.O. depletion were monitored
in a flowing water channel apparatus designed to allow the passage
of a 'plug' of low oxygen water. In an experimental test, 10
individuals were confined to the upstream half of the channel where,
under bright illumination they would bury themselves in sand covering
the bottom. The plug of water of desired D.O. content was then
flushed over the shrimps when some or all would emerge from the sand
and migrate to the downstream half after an intervening divider was
removed. The number migrating by the time the plug had reached the
mid-point of the channel was expressed as a percentage of the total
and represented the response at the level of D.O. tested. A control
'run' was carried out for each population of shrimps prior to the
experiment and the data analysed by standard log-probit methods to
obtain the median effective concentration (EC50) or avoidance threshold. Experiments were carried out at normal and reduced salinities
and at room temperatures, close to 19°C.

     Due to dilution at the source of supply, normal sea water in all
experiments was not of oceanic character but varied between 27‰
and 33‰ salinity.

     Further details of experimental methods and data analysis are
given by Sedgwick (1978).

RESULTS

Salinity and Hypoxic Tolerance

   Salinity tolerance. Following acclimation to a salinity of
27.5‰ , C.vulgaris was found to be capable of surviving severe
reductions in environmental salinity, the LC50 after 72 hours
exposure being estimated at 4.95‰ (Fig. 1) and the zone of resistance, based on 95% mortality and survival lying between 3 and 9‰ .
Acute mortality was rapid, however, and complete within 20 hours.
When shrimp were acclimated to 12.8‰ there was a significant shift
in LC50 to 3.05‰ but also a change in slope function of the line so
that the resistance zone then extended to less than 1.0‰ . This
result would indicate that pre-exposure to medium dilution would
confer some survival advantage under further dilution, however,
rather higher mortalities during holding conditions (15%) may have
influenced the result to a degree. Nonetheless, in real terms the
effect was small.

   Hypoxic tolerance. The response to hypoxia at $20^{o}C$ was extremely
sharp and consequently the resistance zone was narrow, 12-22% air
saturation ($0.9-1.7$ $mgO_2 \ell^{-1}$) for 99% mortality/survival (Fig. 2),
with the LC50 estimated at 17.5% air saturation (1.36 $mgO_2 \ell^{-1}$). Few
of the experimental points, therefore, fell within the body of the
graph. In contrast, the rate of mortality was lower than in the
case of osmotic stress and acute mortality was complete slightly
beyond 48 hours (Fig. 3).

   The effect of osmotic and hypoxic stress. The combined influence
of reduced salinity and D.O. on survival was investigated by means
of a factorial experiment with three levels of each parameter. A
second degree polynomial was fitted to the logarithms of resistance
times and the generated surface is shown in Fig. 4.

   The equation of the surface was given by:

$$Y = 2.9666 + 0.2657x_1 + 0.3092x_2 + 0.0224x_1^2 - 0.2296x_2^2 + 0.1204x_1x_2$$

where

$Y = Log_{10}$ resistance time in minutes.

x_1 = D.O. variable, percentage air saturation.

x_2 = salinity variable, parts per thousand.

   The contours or isopleths locate points of equal response,
expressed as a percentage of the maximum response, in this case, 48
hours resistance time. Two distinct areas may be distinguished,
separated arbitrarily by the 30% isopleth. The first occupies a

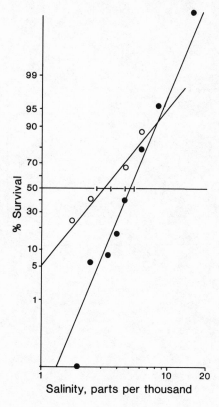

Figure 1. The tolerance of C.vulgaris to reduced salinity at 20°C.
Closed circles – acclimation to 27.5‰. Open circles – acclimation
to 12.8‰. Bars indicate 95% confidence intervals of LC50.

broad region of low salinity and D.O. conditions in which resistance
time was short and varied little with either parameter. In the other
the response increased rapidly with D.O. and salinity as the stress
was progressively lifted. At salinity levels above approximately
20‰ the more horizontal character of the contours indicates that
D.O. was the effective variable while further dilution caused the
lines to bend diagonally to reveal a region of mixed lethal influence. Here the same resistance time can be brought about by a wide
range of different conditions. As the isopleths approach the
vertical, salinity becomes the dominant parameter. Compression of
the contours is indicative of a rapid change in the rate of response
and this is evident from the surface over the middle range of salinity, approximately 10–18‰ and upwards of 15.0% air saturation.
Analysis of variance of partitioned data (Table 1) showed that within this region the variables interacted in affecting the response,
i.e. resistance times were shorter than expected from simple addition
of the component effects.

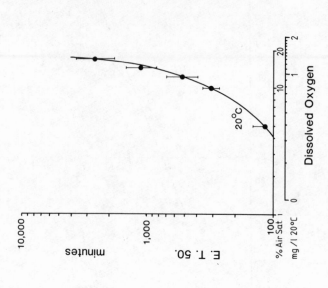

Figure 3. Survival curve for C.vulgaris exposed to low dissolved oxygen at 27.5‰ salinity. Vertical lines indicate 95% confidence intervals of resistance time.

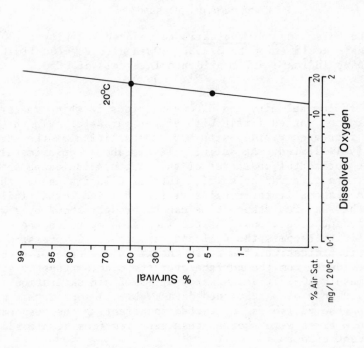

Figure 2. The tolerance of C.vulgaris to low
dissolved oxygen at 27.5‰ salinity.

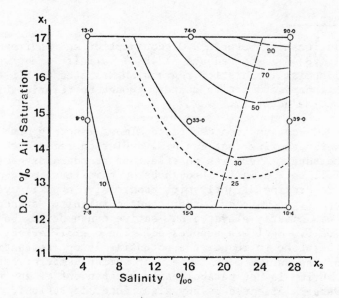

Figure 4. Fitted response surface of resistance times for C.vulgaris exposed to low dissolved oxygen and reduced salinity at 20°C. Open circles indicate experimental points with observed response. Dashed line indicates main axis of surface.

Table 1   Analysis of Variance of Partitioned Data from Factorial
          Experiment

Source of Variation	Sum of squares	Degrees of freedom	Mean Square	F-value	Significance
Salinity	0.5745	1	0.5745	140.1	P <0.001
Dissolved oxygen	0.3470	2	0.1735	42.3	P <0.01
Interaction	0.1879	2	0.0940	22.9	P <0.01
Residual	0.0249	6	0.0041		
Total	1.1342	11			

Response   = $\log_{10}$ resistance time in minutes

Salinity range = 8.1-16.7‰.

D.O. range   = 12.4-17.5% air saturation.

Oxygen Consumption

The relationships between D.O. consumption and environmental D.O. under normal (28.4‰) and reduced (12.8‰) salinity are given in Fig. 5. Polynomial (quadratic) regression was used to fit the curves for standard and active states, though no relationships were sought at routine activity.

Active D.O. consumption decreased almost linearly with D.O. in normal sea water showing that any D.O. depletion reduced the potential energy expenditure. There was no effect on standard oxygen uptake until D.O. fell to 2.0 mg/1 although below this level oxygen consumption tended to decline slightly with progressive hypoxia even though potential capacity remained. Under reduced salinity, standard oxygen consumption was clearly elevated but active consumption above ambient D.O. levels of 6.0 mg/1 was suppressed. The overall effect of osmotic stress was therefore to reduce the potential 'scope for activity' (Fig. 6) over the entire D.O. range. At summer temperatures consequently, the shrimp is not metabolically independent of environmental D.O. and adversely affected by any significant depletion.

Avoidance Behaviour

Observation of the movements of the shrimp during these experiments revealed that emergence and migration, when they occurred, were rapid, keeping pace with the passage of the advancing plug of low oxygen water. The median avoidance threshold was estimated for normal sea water (32.2‰) at 15.3% air saturation (1.13 mg/1) and shifted to 19.4% air saturation (1.68 mg/1) following dilution to 15.3‰

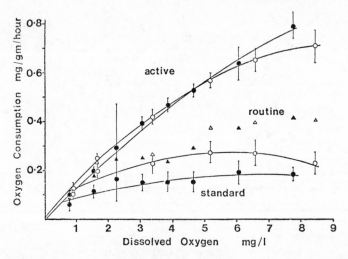

Figure 5. The effect of dissolved oxygen on oxygen consumptions of C.vulgaris at 20°C. Closed symbols – 28.4‰ salinity. Open symbols – 12.7‰ salinity. Vertical lines are 95% confidence intervals of mean dissolved oxygen consumption.

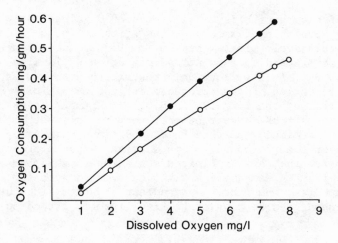

Figure 6. The effect of dissolved oxygen on 'Scope for activity' in C.vulgaris at 20°C. Closed circles - 28.4‰ salinity. Open circles - 12.7‰ salinity.

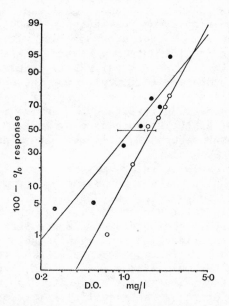

Figure 7. The behavioural response of C.vulgaris to low dissolved oxygen. Mean temperature 19.5°C. Closed circles - 32‰ salinity. Open circles - 15.3‰ salinity. Horizontal lines indicate 95% confidence interval of EC50 values.

salinity. The relationship shown in Fig. 7 were plotted on a D.O. axis expressed in terms of oxygen concentration to accommodate a mean temperature difference of 0.6°C between the two experiments.

Location of the thresholds close to the LC50 values for D.O. strongly
implicates developing dyspnoea as the stimulus for movement and indicates a relatively limited capability for escape. Since the energy
available for expenditure in active migration would also be minimal
at such low D.O. levels, then only restricted progress may be envisa-
ged. The elevation of the threshold at reduced salinity appears
simply to reflect the increase in D.O. requirement measured previously
under osmotic stress. Nevertheless, very rapidly lethal D.O. levels
clearly were avoided, the threshold conditions permitting resistance
times greater than 24 hours. Some survival advantage could be gained, therefore, from even a limited migration if this behaviour is
reproduced in the Thames estuary. Such evasive action may be particularly expedient when the D.O. gradient is steep since only short
distances need then be covered to significantly extend survival times.

DISCUSSION

    Though well able to withstand very severe D.O. depletion of its
environment in sea water, C.vulgaris appears to encounter particular
survival difficulties when hypoxia is experienced simultaneously with
relatively moderate salinity reduction. The basis of the interactive
effect may well be found in the additional energy expenditure associated with osmotic regulation resulting in the elevation of standard
oxygen consumption and avoidance thresholds. It has been argued,
however, that such requirements may be small (Flemister and Flemister,
1951; Potts, 1954; Thabrew et al.,1971). Nevertheless, the progressive loss of 'scope for activity', with increasing hypoxia, exa-
cerbated by osmotic stress must seriously diminish the effectiveness
of the shrimp in asserting itself in the environment. The efficiency
of such activities as the capture of prey and escape from predators
must be considerably reduced in polluted estuarine conditions.

    Although avoidance behaviour may lead to short term modifications
in spatial distribution, the penetration of shrimp populations into
the Thames estuary could be broadly estimated from the results of
osmotic and hypoxic tolerance studies. The application of survival
data to the estuary is confounded by the regular tidal variations
in water quality experienced by an organism holding station. A
simplification can be effected if it is assumed that the overall impact approximates the mean of the fluctuation, i.e. the conditions
existing at mid-tide. The potential penetration may then be estimated
from the static picture of water quality so produced by relating this
with the response surface of resistance times in Fig. 4. A less
precise but more practical estimate of shrimp distribution may be
obtained by considering the more extreme limits of tolerance. The
salinity and D.O. conditions which dictate a resistance time of 12
hours (25% response in Fig. 4) would destroy 50% of the population
within one tidal cycle. The physical position where such conditions
pertain at high water slack would indicate a location where few
shrimp could survive long since, as the tide ebbed, D.O. and salinity

would decline further. Alternatively, few mortalities might be
anticipated where water quality conditions at low water permit a
resistance time of 48 hours. Between these positions exists a region
in which the shrimp populations may be continually redistributed by
the ebb flow bringing the D.O. sag seawards and by instinctive
pressure to follow the temperature gradient further into the estuary.
Since avoidance thresholds approximate to 24 hour resistance time
conditions (50% response), this isopleth may indicate the landward
penetration limits. Although avoidance behaviour may not necessarily
enable shrimp to escape water quality changes perpetrated by the
tidal flow, active migrations might keep pace with variations in
the form of the D.O. sag resulting from fluctuations in organic
loading. These changes frequently require 2-3 days to alter significantly mean D.O. levels measured at a single point along the length
of the estuary (Huddart, 1971).

REFERENCES

Birtwell, I.K. 1972. Ecophysiological aspects of tubificids in the
    Thames Estuary. Ph.D. Thesis. University of London.
Box, G.E.P. 1956. The determination of optimum conditions. In:
    O.L. Davies (Ed). Design and Analysis of Industrial Experiments.
    Oliver and Boyd, London. pp. 495-578.
Flemister, L.J. and Flemister, S.C. 1951. Chloride ion regulation
    and oxygen consumption in the crab Ocypode albicans. Biol.
    Bull., 101: 259-273.
Huddart, R. 1971. Some aspects of the ecology of the Thames
    estuary in relation to pollution. Ph.D. Thesis. University of
    London.
Litchfield, J.T. 1949. A method for rapid graphic solution of time
    per cent effect curves. J. Pharmac. exp. Ther., 97: 399-408.
Litchfield, J.T. and Wilcoxon, F. 1949. A simplified method for
    evaluating dose effect experiments. J. Pharmac. exp. Ther.,
    96: 99-113.
Potts, W.T.W. 1954. The energetics of osmotic regulation in brackish
    and freshwater animals. J. exp. Biol., 31: 618-630.
Sedgwick, R.W. 1978. Some aspects of the ecology and physiology of
    nekton in the Thames estuary with special reference to the
    shrimp, Crangon vulgaris. Ph.D. Thesis. University of London.
Thabrew, M.I., Munday, K.A. and Poat, P.C. 1971. Cation transport
    and metabolism as a function of salinity in the excised gill
    of Carcinus maenus. Comp. Biochem. Physiol., 39: 699-708.

THE PENETRATION OF BRACKISH-WATER BY THE ECHINODERMATA

Richard M. Pagett

Department of Earth Sciences
The University
Leeds, LS2 9JT

INTRODUCTION

Various classifications of brackish-water based on physical
considerations have been proposed (these include Redeke, 1922, 1933;
Välikangas, 1926; Remane, 1934, 1940; Ekman, 1953). Biological
considerations have also been applied (Heiden, 1900; Välikangas,
1933). The former author graded regions of differing salinity on
the basis of particular assemblages of diatoms. There have been
recent reviews concerning the classification of brackish-water
(Symposium on the Classification of Brackish Waters, 1959; Remane
and Schlieper, 1971). Following Kinne (1964c), it is probably
sufficient for the purpose of this presentation to term any body
of water with a salinity lying between 0.5‰ and 30‰ as brackish-
water.

The problems of penetration into brackish-water by marine
invertebrates have been the subject of much research in the past,
recent reviews include (Prosser and Brown, 1961; Kinne, 1964b;
Remane and Schlieper, 1971). However, most of this work has
concentrated upon phyla other than the Echinodermata despite the
number of species of this group which appear either to live wholly
or partly in water with a depressed salinity (see Binyon, 1972b).
Where echinoderms have been the subject of such physiological studies
(ASTEROIDEA - Loosanoff, 1942, 1945; Schlieper, 1957; Binyon,
1961, 1962, 1972a, 1976, 1978, 1980; Pearse, 1967; Ellington and
Lawrence, 1974; Stickle and Ahokas, 1974. ECHINOIDEA - Gezelius,
1963; Giese and Farmanfarmaian, 1963; Lange, 1964; Stickle and
Ahokas, 1974; Lawrence, 1975. HOLOTHUROIDEA - Koizumi, 1932, 1935;
Stickle and Ahokas, 1974) the Ophiuroidea appear to have received
little attention (Pagett, 1980a, 1980b). It is probably timely to

135

summarise the physiological problems of survival, encountered by
marine invertebrate species during euryhaline penetration and to give
some indication of previous work which has gone some way towards
describing the strategies employed by those echinoderms that do
occur, to a lesser or greater degree, in waters of reduced salinity.
The occurrence of Ophiura albida in the brackish-water Loch Etive
will be discussed in order to illustrate the physiological plasticity
that some ophiuroids have for withstanding temporary or permanent
salinity depression.

TOLERATING DEPRESSED SALINITIES

The Problems

The basis of the physiological problems facing brackish-water
organisms is the effect which the lowered salt content of the environment has upon the animal. The integument of a typical soft-bodied
marine invertebrate is permeable to water (Bethe, 1930). The echinoderm integument is no exception. Early work by Frédéricq, (1901),
Macallum (1903) and Botazzi (1908) had shown that the surfaces of
marine invertebrates act as semi-permeable membranes. However,
Bethe (1934), using the opisthobranch mollusc Aplysia, convincingly
demonstrated that this was incorrect and that ions could pass
through the integumental membranes.

Koizumi (1932, 1935) examined the permeability of the body
surface of the holothurian Caudina chilensis and reported that
univalent ions permeate faster than divalent ions. This selection
of univalent ions is probably of general application (Remane, 1939).
Remane and Schlieper (1971) further suggest that it is unlikely
that permeability of the integument of species living in waters of
reduced salinity should be lower for only single ions. They further
propose that the rate of penetration of one ion should permit general
conclusions to be made regarding ion-permeability and, perhaps, also
for the passage of water. Krogh (1939) too, considers that the integumental permeability for ions and for water may be related.
Employing sodium iodide, Ussing (1934) showed that stenohaline species
are more permeable than euryhaline species. Euryhaline species and
those that are largely independent of the external salinity are
the least permeable.

There have been few permeability studies conducted on echinoderms. Subsequent to the studies of Koizumi (1935) the majority of
these studies concern asteroids (Kowalski, 1955; Schlieper, 1957;
Binyon, 1961, 1962, 1972a, 1976, 1980). The latter author investigated the permeability to water of Asterias rubens from the North
Sea. It was demonstrated that this species is permeable to water
and that this permeability remained unchanged during the year.

When a typically marine echinoderm, isosmotic with full strength

sea water, is placed in diluted sea water, water passes through the
integument into the animal by osmosis, diluting the body fluids until
a new, and lower, osmotic pressure equilibrium is achieved. It is
likely that ions in the body fluid diffuse out of the animal down
the concentration gradient between the body fluids and the medium
and that this would also lower the osmotic pressure of the body
fluid (Bethe, 1930). So, the establishment of a new equilibrium
is probably due to both water movement into the animal and to salt
movement in the opposite direction.

Some Solutions

Kinne (1966) proposes that there are various compensatory mechanisms available to a marine invertebrate should the ambient salinity
be depressed:

(1) Escape

(2) Reduction of contact with the lowered salinity

(3) Regulation

(4) Adaptation

The ability to escape would require a reasonably efficient means
of detecting lowered salinity. Pearce (1936) remarks that detection
of slight changes in salinity by many aquatic invertebrates is known.
Loosanoff (1945) suggests that asteroids, in general, and Asterias
forbesi in particular, are capable of such sensory ability.

Reduction of contact by echinoderms, when faced with a lowered
ambient salinity, may be effected by, possibly three methods. Firstly
the production of mucus by Asterias rubens, found at the head of
Loch Etive, may be a response to lowered salinity (Binyon, 1976).
Secondly, Stancyk (1975) proposes that Ophiothrix angulata, from
the Cedar Key region of Florida, has a shortened pelagic life which
may enable it to avoid extremes of salinity. Finally, Turner (1974)
proposes that selective post-metamorphic growth of the arms in the
ophiuroid Ophiophragmus filograneus may permit an early descent of
the disc into the substratum, and thus remove it from contact with
water of depressed salinity present in the surface layers. It has
been recognized that the ability to burrow in marine invertebrates
may enable them to avoid extremes of salinity (Topping and Fuller,
1942; Kinne, 1966). In echinoderms this method has been suggested
for Ophiophragmus filograneus (Stancyk, S.E., personal communication).

As yet, there does not appear to be any extensive ionic or
osmoregulatory system universal within the Echinodermata. Ellington
and Lawrence (1974) report a degree of volume regulation in the
asteroid Luidia clathrata, which contrasts with the absence of any
such regulation in Asterias rubens (Binyon, 1961; Shumway, 1977).

Earlier work on Asterias rubens (Bethe, 1934; Maloeuf, 1938) con-
flicts with these findings; however, it has been suggested (Binyon,
1961) that this is due to experimental design in the earlier work.
Pearse (1967) reports some weight regulation (via inbibition into
the supradorsal cavity) in the antarctic asteroid, Odontaster
validus, while none was demonstrated by Giese and Farmanfarmaian
(1963) in the echinoid Strongylocentrotus purpuratus. It has
been suggested that there may be an optimal coelomic volume to body
volume ratio which would tend to be maintained (Freeman, 1966;
Binyon, 1972a).

     Some degree of ionic regulation has been described in Asterias
rubens from the North Sea (Binyon, 1962). There appears to be some
accumulation and regulation of calcium ions in the perivisceral
fluid, and of potassium ions in the ambulacral fluids of the water
vascular system. Seck (1958) examined this species from the
Western Baltic Sea and did not report any accumulation of calcium
ions. An enhancement of the potassium ion in the perivisceral
coelomic fluid of the ophiuroid Ophiocomina nigra has been suggested
(Pagett, 1980a). It was found that sodium and chloride levels fell
in parallel with the medium while the potassium concentration was
maintained significantly higher (33%) than that in the environment
down to at least 70% sea water.

     Isosmotic intracellular regulation is a phenomenon whereby cells
do not take up as much water as might be expected by the degree of
dilution of the medium when a soft bodied marine animal is placed
in diluted sea water. It is thought that this is due to regulation
at the cellular level rather than at the tissue level. This may be
by a variation perhaps, in the selective membrane permeability or
shunting of various chemical components e.g. amino acids, in order
to maintain cellular osmotic pressure. A response of this type is
described in Asterias rubens (Jeuniaux et al., 1962; Binyon, 1972a)
in which there was found a smaller than expected rise in tissue
hydration during a lowering of salinity. This is thought to be
indicative of the presence of an isosmotic intracellular regulation.
Such regulation has also been reported for the echinoid
Strongylocentrotus droebachiensis (Lange, 1964) and in the asteroid
Luidia clathrata (Ellington and Lawrence, 1974). In this type of
regulation it is thought that amino acids are of paramount importance.
Marine invertebrates which have a 'free amino acid pool' have been
shown to regulate this pool in response to salinity change (Stephens
and Schinske, 1961). Such responses have been investigated by
several authors (Stevens and Virkar, 1966; Emerson, 1969; Stickle
and Ahokas, 1974; Stickle and Denoux, 1976). It has been demonstrated that the amino acid pool decreases as the ambient salinity
decreases. Stevens and Virkar (1966) proposed that this is due to
the incorporation of amino acids into polypeptides, a more economical
process than excretion. Florkin (1962) has suggested that this
represents an osmoregulatory response in the sense that a decrease

in the size of the pool 'cusions' larger fluctuations in the other cellular constituents. Stevens and Virkar (1966) demonstrated that the ophiuroid Ophiactis arenosa was capable of removing amino acids from very dilute solution in seawater, i.e. uptake may maintain the pool when the latter was being drained due to salinity depression. They suggest that the salinity decrease produces an increase in the incorporation of free amino acids into polypeptides and this causes a decline in the pool. Lange (1964) describes three stages in the isosmotic intracellular regulation of Strongylocentrotus droebachiensis - an isosmotic step, an intermediate step and, finally, a regulatory step where the amino acid pool is utilised.

It is probable, however, that the main strategy employed by echinoderms that penetrate waters of reduced salinity is that of adaptation. This enables the organism to adjust to changes in the critical salinity. Such adaptation is of two main types: genetic and non-genetic (Kinne, 1964c; Schlieper, 1967). The former governs increases in tolerance to fluctuating salinity where the main selection mechanism is acting upon the available genetic material. The latter, non-genetic adaptation, involves quantitative changes in the response mechanism of an individual which are not passed on, as such, to succeeding generations. These changes may lead to significant modifications with respect to lethal limits, activity, metabolism, reproduction and other functions.

Stancyk (1973) studied the ophiuroids and echinoids of the estuary at Cedar Key, Florida. He suggests that up to 75% of the inshore species display physiological, morphological and behavioural adaptations in response to the estuary conditions at Cedar Key. The ophiuroid Ophiothrix angulata relies on high dispersal and the ability to colonise or recolonise disturbed habitats after local extinction due to rapid salinity depression. This is in contrast to two other ophiuroids, Ophiophragmus filograneus and Ophioderma brevispinum, which show adaptations that help the young stages avoid such stress conditions (Stancyk, 1973). The euryhalinity of Ophiophragmus filograneus has been further investigated by Turner (1980) and he provides the first documented case of an endemic brackish water echinoderm.

It has been proposed that egg size in ophiuroids correlates with development (Schoener, 1972) and that direct egg development would remove larvae from possible salinity depression in the surface waters (Pearce, 1969; Stancyk, 1973). The ionic composition of sea water affects sperm motility and longevity. Timourian and Watchmaker (1970) used the echinoderm Lytechinus pictus and demonstrated that the spermatozoa are capable of adapting to a reduction in the salt content of the ambient medium and, therefore, are able to fertilize eggs in lowered salinities.

PHYSIOLOGICAL RACES?

Schlieper (1957) compared various features; percentage water
and ash content, and rates of metabolism, in some of the soft tissues
of Asterias rubens from the Baltic Sea and from the North Sea. He
concluded that some of the differences may have arisen as a response
to lowered salinity. Although Asterias rubens is able to reproduce
in the depressed salinities (15‰) of the Kieler Fjord, in the
Western Baltic Sea, he is not of the opinion that this is conclusive
of a physiological race. He suggests that all the differences
between Baltic and North Sea animals could be derived by a more or
less lengthy process of adaptation. It should be mentioned that
Asterias rubens is unable to breed at the extremes of its range in
the Baltic Sea (Segerstråle, 1957, citing Brattström, 1941). Off
the Swedish coast, the echinoid Psammechinus miliaris exists as
two distinct forms, 'S' and 'Z', which occur at different depths
and thus live in different salinities. Gezelius (1963) attempted
to characterise the optimum salinity regimes of the two forms by
using the different cleavage rates for egg development. He demonstrated that adaptation to depressed salinity involved changes in
tolerance and cleavage rate. It was possible to cross-adapt these
forms to the salinity range of the other form. The differences
between these two forms should not be considered as permanent
modifications of hereditary characteristics since they appear to
be morphological forms which have developed in response to the
various regimes of salinity, light and temperature during metamorphosis and early development (Gezelius, 1963).

In Loch Etive, a brackish water loch on the fjordic coastline
of western Scotland (see Fig. 1), there are two species of ophiuroids
Ophiura albida and Amphiura chiajei, which are living in lowered
salinities (28‰ or less). Sill development within the loch
attentuates the amplitude of the tidal range and delays the tidal
phase in comparison with the tides impinging upon the adjacent open
coast. These two effects upon the tidal range, together with a
relatively large fresh water input (Loch Etive has a catchment area
of some 1400 km^2) have lowered the salinity of the loch. The hydrography of Loch Etive has been described by Gage (1972a, 1974) and by
Edwards and Edelston (1977). Similarities between this hydrography
and that of the Baltic Sea have been discussed by Pagett (1978).

Pagett (1980b) compared the salinity tolerance of O.albida and
A.chiajei from Loch Etive with O.albida from Loch Creran (an adjacent
marine sea loch) and with Ophiocomina nigra from Plymouth (see
Figure 2). It is suggested the O.nigra has very little tolerance to
a reduction in its ambient salinity. O.albida from Loch Creran show
some tolerance, while O.albida from Loch Etive are the most tolerant.
As one progresses up the loch from the seaward end towards the head
of the loch, the salinity decreases and the tolerance of O.albida
to reduced salinities increases. A.chiajei also shows increases in

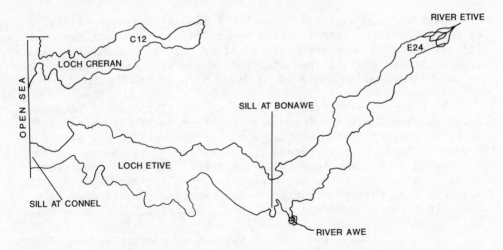

Figure 1. Map of Loch Creran and Loch Etive (not to scale). This map shows the approximate location of sills and sampling stations. The designations of the sampling stations follows Gage (1972).

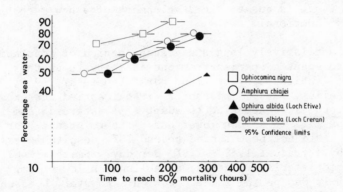

Figure 2. Comparison of salinity tolerances between Ophiocomina nigra, Amphiura chiajei (Loch Etive: E24) Ophiura albida (Loch Etive: E24) Ophiura albida (Loch Creran).

tolerance though not to the same degree. Pagett (1980b) considers that this is because the burrowing, sedentary habit of A.chiajei removes it from the salinities encountered by the surface dwelling O.albida. At the head of the loch (E24) the salinity of the bottom water is about 24‰ but, during two days of continuous rain in October 1977, this was observed to fall to 18‰, a substantial and rapid decrease in salinity. Specimens of O.albida were collected by SCUBA and maintained in these salinities for the next four months. The observations were discontinued after that period since regeneration of broken arm tips had occurred, suggesting that there had been no long term deleterious effects due to the low salinities.

That such a significant difference in tolerance to reduced salinity should exist between O.albida from Lochs Creran and Etive leads to the consideration that such differences are due either to a more or less lengthy process of adaptation or to the selection of favourable mechanisms leading to the production of physiological races. Though physiological races are known for many species, the occurrence of this phenomenon has not yet been demonstrated convincingly in those echinoderms which have penetrated brackish water.

Gezelius (1963), quoting the work of Horstadius (1925) on the echinoid Paracentrotus lividus, notes that adaptation to different salinity regimes has much the same characteristics as the adaptation to new temperature ranges, involving changes in tolerance and cleavage rate of fertilised eggs. He suggests that two processes facilitate such adaptations and that they are not too distantly related physiologically. Beadle (1943) postulates that the extension of the tolerance range is due to the development of a new mechanism not originally functional.

That a relatively long and gradual period of adaptation is required to adapt a species from one salinity to another, lower, salinity is indicated by previous work (Loosanoff, 1945; Schlieper, 1957; Gezelius, 1963). Loosanoff (1945) carried out experiments on Asterias forbesi, attempting to adapt this asteroid from its usual salinity of 27‰ to a much lower one. The tolerance, for 100% survival, could not be extended beyond 18‰ despite lowering the salinity by a gradual 1.5‰ - 2.9‰ per day. Smith (1940), using the asteroid Asterias vulgaris, established a lower limit of 14‰ within a 6-10 day period for adaptation. However, Loosanoff (1945) points out that such periods are likely to be too short. As evidence, he cites his own work in which he maintained Asterias forbesi at 16‰ for a similar period of adaptation as Smith. It was noted that deleterious effects of this lower salinity only became evident after two months. Gezelius (1963) states that about 50 days are required to achieve a complete adaptation between the 'Z' (shallow) and 'S' (deep) forms of the echinoid Psammechinus miliaris. Schlieper (1957) found that Asterias rubens can be transferred between the Baltic and North Sea salinities by a progressive adaptation over several weeks.

Much work, mostly on the Asteroidea and the Echinoidea, has been concerned with the tolerance to differing salinities in the assessment of whether two populations of a species are separate physiological entities. Schlieper (1957) states that there is no

indication that the two populations of Asterias rubens from the
Baltic and North Sea are genetically different races. Gezelius
(1963) kept 'S' and 'Z' forms of P.miliaris in salinities of the
other form for more than a year and noted that there were no major
morphological changes in habitus. He comments that the occurrence
of morphological transformation between the two forms "must be rare".
Borei and Werstedt (1935) state that it is unlikely that the habitat
of the 'S' is populated by individuals of the 'Z' form.

A CROSS ADAPTATION EXPERIMENT BETWEEN OPHIURA ALBIDA FROM LOCHS
CRERAN (MARINE) AND ETIVE (BRACKISH-WATER)

      To investigate further the nature of the population of Ophiura
albida in Loch Etive, in terms of whether these individuals are
tending towards a physiological isolation distinct from those which
are living off the west coast of Scotland and in Loch Creran, a
cross adaptation experiment was carried out. Twenty animals from
Loch Creran, where full salinity conditions obtain (33-34‰), were
used in an attempt to assess their ability to tolerate and survive
an extensive but gradual reduction in the ambient salinity. Since
this loch is fully tidal (Gage, 1972a) the salinity regime is
considered to be equivalent to that off the southern tip of Lismore
(Long. 5°36'0"W, Lat. 56°27'30"N) where the nearest population of
O.albida from fully marine conditions occurs (Pearson, T. personal
communication). To effect the adaptation, the animals were placed
in a large stock tank filled with sea water of the appropriate
initial salinity (33‰) and aerated by airstones. Lining the base
of the tank was various detrital material obtained from Loch Creran
by trawling. Animals for adaptation to full salinity conditions
were obtained from using SCUBA. Twenty animals from this station
(24.4‰) were placed, as above, in a container with the appropriate
initial salinity. Two 20 litre bowls containing Loch Creran animals
and Loch Etive animals were set up also and acted as controls, there
being no salinity change in these bowls. The reduction in salinity
was effected by the addition of glass distilled water producing a
decrease in salinity of 1.4‰ per day in the experiment with the
Loch Creran animals. In the experiment with the animals from station
E24 in Loch Etive the salinity was raised from 24.4‰ to 33‰ by
the process of evaporation at a rate of 1.0‰ per day.

      In falling salinities, the animals from Loch Creran soon began
to show signs of the stress such as curling of arms, loss of colour
and a lower level of activity. They began to die at 26‰ (the
eleventh day of the experiment) and were all dead after a further
8 days at 24.9‰. The O.albida from station E24 at the head of
Loch Etive withstood the experimental increase in salinity and were
maintained at the final salinity of 33‰, which was reached in 9
days, for a further 5 months to ensure that there were no long
term morphological deleterious effects (see Loosanoff, 1945). After
5 months, extensive regeneration of the distal segments of the arms

had occurred and it was considered that the animals had adapted to
the new salinity.

There were no obvious morphological differences between these
animals and those which normally live in salinities of 34‰. Furthermore, differences in colour and level of activity were not apparent
between these newly adapted animals and those from Loch Creran
suggesting that these animals had not suffered any apparently deleterious effects due to the increased salinity. It was not possible,
however, to test whether these animals were able to reproduce in
these new salinities. The ultimate test of a physiological race is
the production of viable froms from the cross fertilisation of
individuals of both 'entities' (which, in this case, are O.albida
from Lochs Creran and Etive) since an inherent feature of speciation
is that of reproductive isolation (Lack, 1947). It is not even
clear whether the O.albida population at E24, the station with the
greatest salinity fluctuation and depression, is reproductively
viable or whether the population relies upon larval and/or adult
recruitment from the lower reaches of the loch or from outside the
loch.

The presence of a brackish water layer on the surface waters of
Loch Etive would seem to be an osmotic and ionic barrier to larvae.
To avoid such conditions it is possible that the larvae might travel
in subsurface currents where the salinity is presumably more favourable. Although Fell (1948) reports that Ophiura albida has an eight-
armed larva (prior to metamorphosis), which indicates that it has a
fairly long larval stage and Thorson (1957) has reported that some
members of the genus Ophiura can delay metamorphosis by as much as
3 days, it is not considered that this would be sufficiently long
enough to allow significant penetration of the loch. Also, the
sills would tend to reflect the water mass of subsurface currents.
It seems likely that only surface waters, and some subsurface waters
during 'renewals', would flow over the sills up towards the inner
reaches of the loch. The factors governing the assumed absence of
larvae from the surface waters presumably would apply also to
potential egg transport within the surface waters. If recruitment
via larvae and/or eggs is impractical, then this would imply that
adult penetration is necessary.

For adult recruitment to be a viable proposition it would be
necessary to envisage a large contingent of adults moving over the
sill at Connel into the lower basin and thence into the upper basin
after surmounting the sill at Bonawe. In both instances it would
require large numbers of animals moving to initially unfavourable
conditions. Not only that, the sills would have to be traversed
regularly. If the lower basin animals do not reproduce then adults
would have to be recruited from the nearest O.albida population.
The nearest population is about 12 km from the entrance narrows of
Loch Etive, which would seem to preclude adult recruitment. It would

also seem unlikely that larvae delay metamorphosis long enough to
permit recruitment from this distance. It has been estimated that
a 'packet' of water from this area would take two or three months
to reach the Bonawe narrows (Edwards, A. personal communication).

To assess the occurrence of reproduction of Ophiura albida within the confines of the loch, plankton hauls were taken at the end
of July 1977 when it was considered, with reference to known breeding populations of this species around Britain, that ophioplutei
should be present in the plankton (Tyler, P.A. personal communication). It was not possible to follow the reproductive cycle histo-
logically. Binyon (personal communication) examined the gonads of
Asterias rubens from station E24 and found them ripe and ready to
discharge gametes into the water. He also observed that the eggs
showed germinal vesicle breakdown. Oblique plankton tows were taken
at stations E2, E11 and E24. To allow for the possibility of
ophioplutei occurring at the level just below the halocline a
horizontal tow was made at station E24. No echinoderm larvae were
found. This is not conclusive evidence of a non-breeding population
of Ophiura albida because it is possible that trawling was carried
out at the wrong time. It has been demonstrated that in some echinoderms, for example Ophiocomina nigra, that spawning is confined to
the production of discrete packets of gametes over a very short
period of time (Gorzula, 1976).

The presence of small specimens (1.0mm to 1.5mm in diameter
across the disc) of Ophiura albida in the trawl net at station E24
during April 1977 suggests either, that the breeding season was
extremely early that year, or, that the animals were derived from
the spat of the previous year. It is known that in some echinoderms,
e.g. Asterias rubens, the maturation of the gonads is slower in
brackish water (Schlieper, 1957). From the newly metamorphosed size
and with reference to reproductive studies of O.albida off the Gower
Peninsula, South Wales (Tyler, 1976) it seems likely that these small
brittlestars were spawned in the summer of 1976 and subsequently
overwintered. It seems unlikely that these small brittlestars, if
spawned in the lower basin, could move over the sill at Bonawe into
the upper basin and thence to station E24 at the head of the loch.
Assuming that as larvae they could not tolerate the salinities of
the brackish water layer, it seems plausible that the larvae are
derived from animals within the upper basin. Furthermore, on the
basis of likely rates of movement (Romanes and Ewart, 1881; Galtsoff
and Loosanoff, 1939; Smith, 1940; Reese, 1966) and improbabilities
of larval transport in the surface layers, as discussed previously,
it is considered that Ophiura albida is capable of reproducing at
the head of Loch Etive. It seems unlikely that these small animals
merely travelled up the loch from, say, stations E11 or E14 since,
if this were so, the young animals with a possibly lower tolerance
to reduced salinities than the adult stages, would be moving into
less favourable conditions which seems implausible for such young

and presumably more physiologically sensitive animals.

If the above is correct and the population of Ophiura albida at E24 are reproductively viable, are they tending towards reproductive isolation from the obviously marine O.albida in Loch Creran? Do these speculations concerning O.albida apply equally to Amphiura chiajei and the asteroid Asterias rubens which also are represented at station E24? It is interesting to note that A.rubens shows a marked morphological distinction between members at E24 and a 'typical' A.rubens from, say, the North Sea. Indeed, it has many features in common with those in the brackish-water Baltic Sea.

CONCLUSION

The penetration of brackish-water by members of the Echinodermata is a well recorded, yet relatively little studied, phenomenon. Mostly the literature holds records of the presence of echinoderms in reduced salinities with little additional investigation as to the geographical possibilities for their occurrence nor how these populations are maintained there, if, indeed they are. The example of Ophiura albida discussed here illustrated the potential of some ophiuroids for penetrating brackish-water. It also illustrates the need for wide ranging studies including not only physiological experiments but also ecological observations. Undoubtedly, the problems of penetration are met by the individual by a number of adaptations on different levels. In most cases biochemical data are wanting. Time-series ecological sampling is necessary to determine whether or not a particular individual in brackish-water is reproductively viable or does its population rely upon recruitment from outside the brackish area, that is to say, is it a truly brackish-water species?

ACKNOWLEDGEMENTS

I would like to acknowledge the logistic support from the Scottish Marine Biological Association. I would also like to thank Dr. E.J. Binyon for his helpful remarks during the course of this work.   I acknowledge the permission of the Controller, H.M. Stationery Office and of the Hydrographer of the Navy for permission to base Figure 1 on portions of British Admiralty Charts 2814A and 2814B. This research was financed by the Science Research Council (Grant Number: B 75/0147).

REFERENCES

Beadle, L.C. (1943). Osmotic regulations and the fauna of inland waters. Biological Reviews 18: 172-183.
Bethe, A. (1930). The permeability of the surface of marine animals. Journal of General Physiology 13: 437-444.
Bethe, A. (1934). Die Salz-und Wasser - Permeabilität der Körperoberflächen verschiedener Seetiere in ihrem gegenseitigen

Verhältnis. Pflügers Archiv fur die gesamte Physiologie des Menschen und der Tiere 234: 629-644.

Binyon, J. (1961). Salinity tolerance and permeability to water of the starfish Asterias rubens L. Journal of the Marine Biological Association UK 41: 161-174.

Binyon, J. (1962). Ionic regulation and mode of adjustment to reduced salinity of the starfish Asterias rubens L. Journal of the Marine Biological Association UK 42: 49-64.

Binyon, J. (1966). Salinity Tolerance and Ionic Regulation. In: Physiology of Echinodermata. ed. by R. Boolootian. Chapter 15, 359-377, Interscience, New York.

Binyon, J. (1972a). The effects of diluted sea water upon podial tissues of the starfish Asterias rubens. Comparative Biochemistry and Physiology 41a: 1-6.

Binyon, J. (1972b). Physiology of Echinoderms. Pergamon Press, Oxford.

Binyon, J. (1976). The effects of reduced salinity upon the starfish Asterias rubens L. together with a special consideration of the integument and its permeability to water. Thalassia Jugoslavica 12: 11-20.

Binyon, J. (1978). Some observations upon the chemical composition of the starfish Asterias rubens L. with particular reference to strontium uptake. Journal of the Marine Biological Association UK 58: 441-449.

Binyon, J. (1980). Osmotic and hydrostatic permeability of the integument of the starfish Asterias rubens L. Journal of the Marine Biological Association UK 60: 627-630.

Borei, H. and Wernsteadt, C. (1935). Zur Okologie un Variation von Psammechinus miliaris. Arkiv for Zoologi 289: 1-15.

Botazzi, F. (1908). Osmotischer Druck und electrische Leitfahigheit der Flussikeiten der einzelligen pflanzlichen und tierischen Organismen. Ergebnisse der Physiologie 7: 161-402.

Edwards, A. and Edelsten, D. (1977). Deep water renewal of Loch Etive: A Three Basin Scottish Fjord. Estuarine and Coastal Marine Science 5: 575-595.

Ekman, S. (1953). Zoogeography of the Sea. London, 1-418.

Ellington, W.R. and Lawrence, J.R. (1974). Coelomic fluid volume regulation and isosmotic intracellular regulation by Luidia clathrata (Echinodermata : Asteroidea) in response to hyposmotic stress. Biological Bulletin 146: 20-31.

Emerson, D.N. (1969). Influence of salinity of ammonia excretion rates and tissue constituents of euryhaline invertebrates. Comparative Biochemistry and Physiology 29: 1115-1133.

Fell, H.B. (1948). Echinoderm embryology and the origin of the chordates. Biological Reviews 23: 81-107.

Florkin, M. (1962). La régulation isosmotique intracellulaire chez les invertébrés marins euryhalins. Bulletin de l'Académie Royale de Belgique 48: 687-694.

Frédéricq, L. (1901). Sur la concentration moleculaire du sang et des tissues chez les animaux aquatiques. Bulletin de l'Académie

r. de medecine de Belgique Classe des sciences 8: 428-454.

Freeman, P.J. (1966). Observations on osmotic relationships in the holothurian Opheodesoma spectabilis. Pacific Science 20: 60-69.

Gage, J.D. (1972a). A preliminary survey of the benthic macrofauna and sediments in Lochs Etive and Creran, sea-lochs along the west coast of Scotland. Journal of the Marine Biological Association UK 52: 237-276.

Gage, J.D. (1974). Shallow water zonation of sea-loch benthos and its relation to hydrographic and other physical features. Journal of the Marine Biological Association UK 54: 223-249.

Gezelius, G. (1963). Adaptation of the sea urchin Psammechinus miliaris to different salinities. Zoologie Bidrag, 35: 329-337.

Giese, A.C. (1966). On the biochemical constitution of some echino-derms. In: Physiology of Echinodermata (Boolootian, R.A. ed). Interscience, New York, 822pp.

Giese, A.C. and Farmanfarmaian, A. (1963). Resistance of the purple sea urchin to osmotic stress. Biological Bulletin of the Marine Biological Laboratory, Woods Hole 124: 182-192.

Gorzula, S. (1976). The Ecology of Ophiocomina nigra in the Firth of Clyde. Ph.D. Thesis, London.

Heiden, H. (1900). Diatomeen des Conventer Sees bei Doberan etc. Milleilungen aus dem GroBherz. Merkl. - Geol Landesanst 10.

Jeuniaux, C. et al. (1962). Régulation osmotique intracellulaire chez Asterias rubens L. Role du glycolle et de la taurine. Cahiers de biologie marine 3: 107-113.

Kinne, O. (1964b). The effects of temperature and salinity on marine and brackish-water animals. Oceanography and Marine Biology Annual Reviews 2: 281-339.

Kinne, O. (1964c). Non-genetic adaptation to temperature and salinity. Helgolander wissenschaftliche Meeresuntersuchungen 9: 443-458.

Kinne, O. (1966). Physiological aspects of animal life in estuaries with special reference to salinity. Netherlands Journal of Sea Research 3: (2) 223-244.

Koizumi, T. (1932). Studies on the exchange and the equilibrium of water and electrolytes in a holothurian, Caudina chilensis. Science Reports of the Research Institutes, Tohoku University Series 4: 259-311.

Koizumi, T. (1935). Studies on the exchange and the equilibrium of water and electrolytes in a holothurian, Caudina chilensis. Science Reports of the Research Institutes, Tohoku University Series 10: 269-275.

Kowalski, R. (1955). Untersuchungen zur Biologie des Seesternes Asterias rubens L. in Brackwasser. Kieler Meeresforsuchungen 11: 201-213.

Krogh, A. (1939). Osmoregulation in Aquatic Animals. Cambridge University Press, Cambridge, 242pp.

Lack, D. (1947). Darwin's Finches. Cambridge.

Lange, R. (1964). The osmotic adjustment in the echinoderm Stronglocentrotus droebachiensis. Comparative Biochemistry and Physiology 13: 205-216.

Lawrence, J.M. (1975). The effect of temperature - salinity combinations on the functional well being of adult Lytechinus variegatus (Lamarck) (Echinodermata, Echinoidea). Journal for Experimental Marine Biology and Ecology 18: 271-275.

Loosanoff, V.L. (1942). Observations on starfish, Asterias forbesi exposed to sea water of reduced salinities. Anatomical Record Abstract 84: 86.

Loosanoff, V.L. (1945). Effects of sea water of reduced salinity upon starfish, Asterias forbesi of Long Island Sound. Transactions of the Connecticut Academy of Arts and Sciences 36: 813-835.

Macallum, A. (1903). On the inorganic composition of the medusae, Aurelia flavidula and Cyanea artica. Journal of Physiology 29: 213-242.

Maloeuf, N.S.R. (1938). Studies on the respiration and osmoregulation of animals. Zeitschrift für vergleichende Physiologie 25: 1-28.

Pagett, R.M. (1978). Some physiological and ecological aspects of the penetration into brackish water by certain members of the Ophiuroidea. Ph.D. Thesis (Unpub.) University of London, 288pp.

Pagett, R.M. (1980a). Distribution of sodium, potassium and chloride in the ophiuroid, Ophiocomina nigra (Abildgaard). Journal of the Marine Biological Association UK 60: 163-170.

Pagett, R.M. (1980b). Tolerance to brackish water by Ophiuroids with special reference to a Scottish sea Loch, Loch Etive. In: Echinoderms: Present and Past (Jangoux, M. ed) Proceedings of the European Colloquium on Echinoderms, Brussels, 3-8 September, 1979. Balkema, A.A., Rotterdam, 223-229.

Pearse, A.S. (1936). The migrations of animals from sea to land. Duke University Press, 1-176.

Pearse, J.S. (1967). Coelomic water volume control in the Antarctic sea star Odontaster validus. Nature 216: 1118-1119.

Pearse, J.S. (1969). Slow developing demersal embryos and larvae of the antarctic sea star Odontaster validus. Marine Biology 3: 110-116.

Prosser, C.L. and Brown, F.A. (1961). Comparative Animal Physiology. 2nd Edition Saunders, Philadelphia, 688p.

Redeke, H.C. (1922a). Biologie der niederländischen Brackwassertypen. Bijdragen tot de dierkunde Amsterdam 22: 239-335.

Redeke, H.C. (1933). Uber den jetzigen Stand unserer Kenntnisse der Flora und Fauna des Brackwassers. Verhandlungen der Internationalen Vereinigung für Limnologie 6: 46-61.

Reese, E.S. (1966). The complex behaviour of Echinoderms. In: The Physiology of Echinodermata, ed. R.A. Boolootian, 157-218.

Remane, A. (1934a). Die Brackwasserfauna. Verhandlungen der Deutschen zoologischen Gesellschaft, 34-74.

Remane, A. (1940). Einfuhrung in die zoologische Ökologie der Nord- und Ostee. In: Tierwelt der Nord-und Ostee, 1a: 1-238.

Remane, A. (1959). Regionale Verschiedenheiten der Lebewesen gegenüber dem Salzgehalt und ihre Bedeutung für die Brackwasser-

Einteilung. Arch. Oceanogr. e Limnol. (Venezia) 11: (Suppl)
    35-46.
Remane, A. and Schlieper, C. (1957). Biology of Brackish Water.
    Die Binnengewasser XXV, Interscience.
Romanes, G.J. and Ewart, J.C. (1881). Observations on the locomotor
    system of Echinodermata. Philosophical Transactions of the
    Royal Society 172: 829-885.
Schlieper, C. (1957). Comparative study of Asterias rubens and
    Mytilus edulis from the North Sea and the Western Baltic.
    Annals Biology 33: 117-127.
Schlieper, C. (1967). Genetic and non-genetic cellular resistance
    adaptation in marine invertebrates. Helgolander wissenschaft-
    liche Meeresuntersuchungen 14: 482-502.
Schoener, A. (1972). Fecundity and possible mode of development of
    some deep-sea ophiuroids. Limnology and Oceanography 17: 193-
    199.
Seck, Ch. (1958). Untersuchungen zür Frage der Tonenregulation bei
    in Brackwasser lebenden Evertebraten. Kieler Meeresforschungen
    13: 220-243.
Segerstråle, S. (1957a). Baltic Sea. Memoirs of the Geological
    Society of America 67: 1-32.
Shumway, S.E. (1977). The effects of fluctuating salinities on four
    species of asteroid echinoderms. Comparative Biochemistry and
    Physiology 58a: 177-179.
Smith, G.F.M. (1940). Factors limiting distribution and size in
    the starfish Asterias forbesi. Journal of the Fisheries Research
    Board of Canada 5: (1) 84-104.
Stancyk, S.E. (1973). Development of Ophiolepis elegans (Echino-
    dermata : Ophiuroidea) and its implications in the estuarine
    environment. Marine Biology 21: 7-12.
Stancyk, S.E. (1975). The life history pattern of Ophiothrix
    angulata (Ophiuroidea). American Zoologist 15: (3) 793 (abst-
    ract).
Stancyk, S.E. and Shaffer, P.L. (1977). The salinity tolerance of
    Ophiothrix angulata (Say) (Echinodermata : Ophiuroidea) in
    latitudinally separate populations. Journal for Experimental
    Marine Biology and Ecology 29: 35-43.
Stephens, G.C. and Schinske, R.A. (1961). Uptake of amino acids by
    marine invertebrates. Limnology and Oceanography 6: 175-181.
Stephens, G.C. and Virkar, R.A. (1966). Uptake or organic material
    by aquatic invertebrates. IV The influence of salinity on the
    uptake of amino acids by the brittlestar, Ophiactis arenosa.
    Biological Bulletin, 35:
Stickle, W.B. and Ahokas, R. (1974). The effects of tidal fluctua-
    tions of salinity on the perivisceral fluid composition of
    several echinoderms. Comparative Biochemistry and Physiology
    47a: 469-476.
Stickle, W.B. and Denoux, G.J. (1976). Effects of in situ tidal
    salinity fluctuations on osmotic and ionic composition of body
    fluid in Southeastern Alaska Rocky Intertidal Fauna. Marine

Biology 37: 125-135.
Symposium on the classification of Brackish waters. (1959). Venice,
    8-14 April 1958, Arch. Oceanogr. Limnol. (Suppl) XI 248pp.
Thomas, L.P. (1961). Distribution and salinity tolerance in the
    amphiurid brittlestar, Ophiophragmus filograneous, (Lyman 1875).
    Bulletin of Marine Science of the Gulf and Caribbean 11: 158-160.
Thorson, G. (1957). Bottom communities (sublittoral or shallow shelf).
    Memoirs of the Geological Society of America 67: 461-534.
Topping, F.L. and Fuller, J.L. (1942). The accommodation of some
    marine invertebrates to reduced osmotic pressures. Biological
    Bulletin of the Marine Biological Laboratory, Woods Hole, 82:
    372-384.
Turner, R.L. (1974). Post-metamorphic growth of the arms in
    Ophiophragmus filograneous (Echinodermata : Ophiuroidea) from
    Tampa Bay, Florida, USA. Marine Biology 24: 273-277.
Turner, R.L. (1980). Salinity tolerance of the brackish-water
    echinoderm Ophiophragmus filograneus (Ophiuroidea). Marine
    Ecology - Progress Series 2: 249-256.
Tyler, P. (1976). The ecology and reproductive biology of the genus
    Ophiura with special reference to the Bristol Channel. Ph.D.
    Thesis Swansea, 247pp.
Ussing, H.H. (1949). Transport of ions across cellular membranes.
    Physiological Reviews, Washington 29: 127-155.
Välikangas, J. (1926). Planktologische Untersuchungen im Hafengebiet
    von Helsingfors. Acta Zoologia Fennica 1: 1-277.
Välikangas, J. (1933). Über die Biologie der Ostee als Brackwasser-
    gebiet. Verhandlungen der Internationalen Vereinigung für
    Limnologie 6: 62-112.

EFFECTS OF CLOSURE OF THE GREVELINGEN ESTUARY ON SURVIVAL AND DEVELOP-

MENT OF MACROZOOBENTHOS

R.H.D. Lambeck

Delta Institute for Hydrobiological Research
Yerseke
The Netherlands

INTRODUCTION

As a consequence of big hydraulic engineering schemes within
the framework of the so-called Delta project, the Grevelingen estuary
(SW Netherlands) has gone through several stages during the last two
decades.

Up until 1964 it was an estuary forming part of the Rhine-Meuse
system (Fig. 1a). Chlorinity values could fluctuate widely, especially in the eastern part (Peelen,1967). In 1964 a secondary dam was
built, disconnecting the direct ties with the rivers mentioned (Fig.
1b). The newly created tidal inlet was, however, still influenced
by brackish water originating from the Haringvliet and flowing around
the island of Goeree-Overflakkee (Fig. 1a) in periods of a high river-
water discharge. In the eastern part, the chlorinity amplitude became
much smaller and, for the whole Grevelingen, the frequency of low
values decreased (Peelen, 1973).

In May 1971 the Grevelingen estuary was cut off from the North
Sea by the completion of a 6 km long dam, which created a lake of
108 km^2 (Fig. 1c). With a water exchange of less than 10% a year,
Lake Grevelingen was nearby a closed system until 1978. In that
year a newly constructed sluice in the North Sea dam was brought
into operation, bringing the chlorinity (that had, due to the yearly
precipitation surplus gradually decreased from 17‰ to 13‰ Cl$^-$),
back to about 16‰ Cl$^-$. More details on the lake and its (changing)
environmental conditions can be found in e.g. Nienhuis (1978).

To a certain extent the closure of the Grevelingen estuary
may be regarded as a biological experiment, in which the impact of

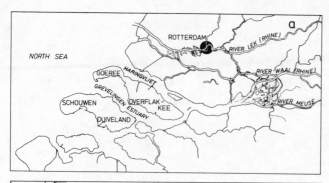

Figure 1.

a) The Grevelingen estuary as a part of the Rhine–Meuse system before 1964.

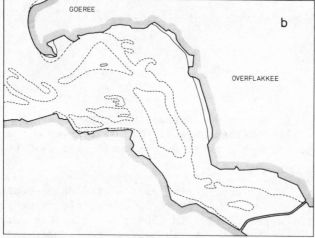

b) The situation in the period 1964–1971. The intertidal areas are indicated by broken lines.

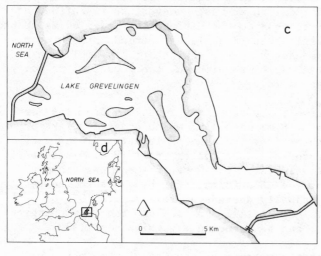

c) Present Lake Grevelingen.

d) The Delta area situated in North-west Europe.

an environmental master factor, the tide, can be studied. By following the long term changes in density and distribution of species one can get more insight into the adaptability or the ecological amplitude of the estuarine fauna (and flora) originally present.

Some aspects of the changes in the softbottom macrofauna up till 1977, concerning the period before the North Sea sluice came into operation, will be briefly outlined and discussed in this paper.

SUMMARY OF RESULTS

A comparison of the species composition within six systematic groups under estuarine (1962 + 1963) and lake conditions (1976 + 1977) is given in Fig. 2. The number of species involved decreased by about 20%. However, only the larger, often migratory and/or predatory epifaunal species, such as crabs, sea spiders and echinoderms were really affected. Other groups, e.g. molluscs and polychaetes, maintained their diversity rather better (of course this does not exclude considerable shifts in relative occurrence).

The majority of the "disappeared" species can be regarded as North Sea ones penetrating into the mouth of the estuary in relatively low numbers. Especially among the polychaetes, a considerable number of "new" species have been found. Most of these are also known from other areas with only weak currents, e.g. coastal lagoons and fjords.

A comparison of the distribution of each species before and after closure, showed that most of them appeared to be more widespread in the lake than in the former estuary.

Mollusc biomass (figures based on a yearly survey with a Van Veen grab) decreased after the closure, but a recovery started in 1975 resulting in a preliminary estimate of about 90 g ash-free dry weight m^{-2} for spring 1977. Estuarine data are restricted to only one year (Wolff and de Wolf, 1977) but the conclusion that recent lake mollusc biomass figures are, at least, not lower than the estuarine ones seems justifiable.

Biomass time series for other systematic groups are lacking, but a study of polychaete density at three permanent stations suggests a mean increase.

DISCUSSION

The occurrence of tides is obviously not a prerequisite for many estuarine macrozoobenthic species. In contrast to the rather stable conditions in the lake, environmental parameters in the estuary could fluctuate widely, which selected for rather "tough" species. Adaptability is the most important strategy for estuarine animals.

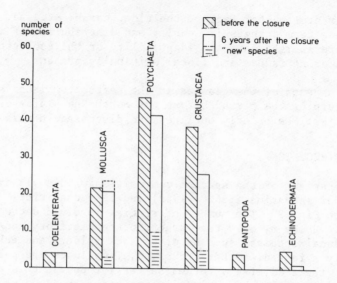

Figure 2. A comparison between the macrozoobenthos composition in the Grevelingen area before and after the closure. Broken line is right mollusc column: three "new" species did live also in the estuary, but were not included in the left column because their then habitat (around the highwater mark) fell outside the research area.

Not withstanding this flexibility, parts of the estuary were not very suitable for benthic life due to strong tidal currents and the resulting sediment movements. Under lake conditions all these former (near) bare areas could be colonized, explaining the extension in the distributional range for many species.

Intertidal areas were among the most productive areas in the estuary (Wolff and de Wolf,1977). The upper intertidal areas became land after fixing the waterlevel (Fig. 1b-c), which wiped out all local benthic life. Mortality did also occur sublittorally as part of a chain-reaction: the decomposition of huge amounts of organic matter ultimately resulted in oxygen depletion, especially in the deeper parts of the lake. Both the above factors may have played a role in the initial biomass decrease.

The recovery delay may be the result of several factors:

a)   a recruitment failure in 1971 as a direct result of the environ-
      mental disturbance initiated by the closure;

b)   the youngest generation of several species, e.g. <u>Macoma balthica</u>
      (Beukema,1973) grows up in the upper intertidal, a habitat that
      became land after the closure, and is dependent on (tidal)
      currents for dispersal to its ultimate place of occurrence.

c)   the predators of molluscs e.g. the starfish <u>Asterias</u> <u>rubens</u>
      and the plaice <u>Pleuronectes</u> <u>platessa</u>, became dependent on a
      much smaller food stock. Moreover, because initially the pop-
      ulation decrease was rather slow, we might expect, taking into
      account the results of de Vlas (1979), a considerable nett
      increase in food consumption of the ageing (locked) Grevelingen
      plaice population. Both factors increased the predation pressure
      on the younger molluscs in particular. A sharp decline in num-
      bers of starfish and plaice started in 1974;

d)   colonization of habitat by "new" species takes time;

e)   several mollusc species have an irregular recruitment, resulting
      in good and bad year-classes. The cockle is an example for this
      phenomenon (Beukema,1979), which means that also natural fluc-
      tuations may have played a role.

The "Grevelingen story" illustrates once more that the recovery
of an ecosystem after a major disturbance is a process that may take
many years.

The data presented here are preliminary in character as studies
into the development of the macrozoobenthos are not yet fully completed. More extensive and integrative publications will appear elsewhere.

REFERENCES

Beukema, J.J. (1973). Migration and secondary spatfall of <u>Macoma</u>
      <u>balthica</u> (L.) in the western part of the Wadden Sea. <u>Neth J.</u>
      <u>Zool.</u> 23: 356-357.
Beukema, J.J. (1979). Biomass and species richness of the macro-
      benthic animals living on a tidal flat area in the Dutch Wadden
      Sea: effects of a severe winter. <u>Neth. J. Sea. Res.</u> 13: 203-
      223.
Nienhuis, P.H. (1978). Lake Grevelingen: a case study of ecosystem
      changes in a closed estuary. <u>Hydrobiol. Bull.</u> 12: 246-259.

Peelen, R. (1967). Isohalines in the Delta area of the rivers Rhine,
    Meuse and Scheldt. Neth. J. Sea Res. 3: 575-597.
Peelen, R. (1973). Variatie in het chloride gehalte van de
    Grevelingen. Delta Inst. for Hydrobiol. Res. unpublished
    report.
de Vlas, J. (1979). Annual food intake by plaice and flounder in a
    tidal flat area in the Dutch Wadden Sea, with special reference
    to consumption of regenerating parts of macrobenthic prey.
    Neth. J. Sea Res. 13: 117-153.
Wolff, W.J. and de Wolf, L. (1977). Biomass and production of zoo-
    benthos in the Grevelingen estuary, The Netherlands. Estuar.
    Coast. Mar. Sci. 5: 1-24.

TIDAL MIGRATION OF PLAICE AND FLOUNDERS AS A FEEDING STRATEGY[*]

W.J. Wolff[†], M.A. Mandos and A.J.J. Sandee

Delta Institute for Hydrobiological Research
Yerseke
The Netherlands

INTRODUCTION

Estuarine tidal flats are characterized by high biomass of benthic invertebrates (see for instance Beukema, 1976; Wolff and de Wolf, 1977). Consequently, these flats constitute rich feeding grounds for large numbers of birds (see for instance Swennen, 1976) and fishes also exploit this abundant source of food. Bückmann (1935) seems to have been the first investigator to draw attention to the possibility of migration of fishes between the tidal flats submerged during high tide and the tidal channels still containing water at low tide. Since then, several others have studied this phenomenon. Kuipers (1973, 1977), for example, describes the tidal migration of juvenile plaice (Pleuronectes platessa) in the Wadden Sea and discusses the results of other authors. Wells et al. (1973) describe intertidal feeding of winterflounder (Pseudopleuronectes americanus) in the Bay of Fundy.

Our data have been collected in the estuaries of the Delta area of the S.W. Netherlands in 1970-73. In this paper we shall try to arrive at an order-of-magnitude quantification of the importance of plaice and flounder (Platichthys flesus) as predators on the tidal flat benthos of the Delta estuaries. Finally, we want to consider the question why some species of fish have adopted the apparently successful feeding (and survival?) strategy of tidal migration.

[*] Contribution nr. 208 of the Delta Institute for Hydrobiological Research.

[†] Present address: Research Institute for Nature Management, P.O. Box 59, Texel, The Netherlands.

METHODS

On the tidal flats we caught the fish with one or more standing nets with a length of 120 m, a height of 30 cm and meshes of 9 cm. This mesh-size catches plaice and flounder only when they are about 18 cm or larger. The bottom rope of the net is weighted with lead, the upper rope is furnished with floats. The net was set at high tide slack water and kept in position by an anchor at each end. At low tide, when the net lay on the dry flat, its contents were inspected and if necessary brought to the laboratory. A few times the nets were set at low tide and taken in at high tide.

In the tidal channels, samples were obtained with a 3 m wide beam trawl with a mesh-size of 1 cm.

For food analysis, stomachs and guts were removed and stored in 4% formalin.

Otolith reading provided the age of the fishes.

Macrofauna was sampled by a 0.005 m^2 corer and sieved through a 1 mm sieve.

AREA

The investigations were carried out on the tidal flats of the Grevelingen estuary near Herkingen in 1970 and 1971 and, after the closure of the former estuary in 1971, on the flats of the Ooster- schelde estuary near the Dortsman channel south of Stavenisse in 1972 and 1973 (Figure 1). Both areas consist of extensive sandy flats with about 2 km distance between high and low water line. The high water boundary of the flats was formed by small areas of salt marsh and the sea wall.

The tidal range was nearly 3 m in the Grevelingen estuary and about 3.5 m in the Oosterschelde estuary.

In both areas, the flats were inhabited by a euhaline-polyhaline benthic flora and fauna (Nienhuis, 1970; Wolff, 1973; Wolff and de Wolf, 1977).

NET EFFICIENCY

An attempt was made to determine the efficiency of the standing nets used by placing three nets in succession. If each net catches a constant proportion p of the fishes meeting the net, then of N fishes meeting the first net p.N will be caught and (1-p).N will meet the second net. Successively $p(1-p).N$, $p(1-p)^2.N$, and $p(1-p)^{k-1}.N$ will be caught in the second, third and k^{th} net, respectively. If the number of fishes caught in each net is called a_1, a_2, a_3 ..... and a_k, then $\ln a_k = \ln p.N + (k-1).\ln(1-p)$.

Regression of $\ln a_k$ against $k-1$ provides the regression coefficient $m = \ln(1-p)$ and the intercept $c = \ln(p.N)$. From this follows that $p = 1-e^m$ and $N = e^c/1-e^m$.

For the data recorded in Table 1 the common regression coefficient becomes $m = 0.52 \pm 0.18$. Hence, $p = 0.40 \pm 0.10$. The estimated values of N are given in Table 1.

However, the data in Table 1 show that compared to net 1, net 2 consistently caught more fish than expected. Therefore, an independent check of the efficiency determined by this method is urgently required. The estimated numbers of fish should be considered as close to a minimum estimate, since the numbers caught in net 1 constitute on average 42% of the total number caught in the three nets together.

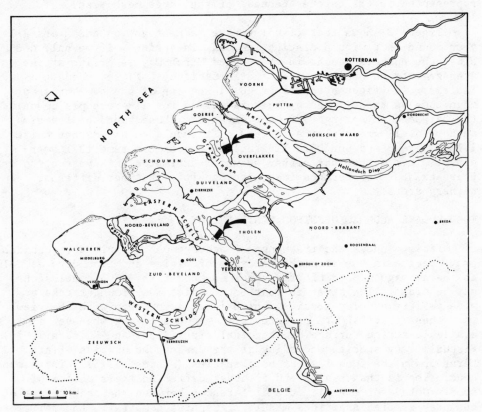

Figure 1. Map of the Delta area in the S.W. Netherlands with the localities where flatfish were sampled for this study.

AGE DISTRIBUTION AND NUMBERS OF FISH ON THE TIDAL FLATS

The nets were set near LW-level at right-angles to the slope of
the tidal flats. The ebb current forced the net into the shape of
an arc with a width of about 100 m. Hence, it was assumed that the
net was catching a proportion of all fishes occurring at high tide
in a rectangle of 100 x the distance between the net and the HW-level
(m^2).

In the Grevelingen estuary this area was about 20,000 m^2, in
the Oosterschelde estuary, about 18,000 m^2.

Because of the large mesh-size the nets only caught the larger
flatfish (Table 2). The catches of plaice largely consist of II-
group individuals; the younger ones probably pass through the net,
whereas the older ones apparently have left the estuaries (Wimpenny,
1953). Therefore, we shall mainly consider II-group plaice in the
following paragraphs. Also the 0- and I-group flounders are hardly
represented in the catches because of the large mesh-size.

Figure 2 shows that flatfish have been caught on the flats all
year round, but with a distinctive peak in spring. If we only consider the numbers caught in the period March-May we arrive at the
average catches and densities (net efficiency 40%) recorded in
Table 3. The figures for II-group plaice are of the same order of
magnitude as those found by Kuipers (1977) for II-group plaice on
the Balgzand in spring 1975, but far below those found by the same
author for II-group plaice in 1973. It should be noted that Kuiper's
1973 II-group and our 1973 II-group belong to the same 1971 year-
class. This means that either we under-estimated this year-class or
there exists a difference in plaice densities between different
nursery areas.

FOOD UPTAKE AND MACROBENTHOS DISTRIBUTION

Table 4 shows the frequency of occurrence of various recognizable
prey species in the stomachs and guts of plaice and flounders as well
as the average number of prey specimens per fish. Two conclusions
may be drawn. The first is that both plaice and flounder take a
wide variety of prey species, which is in accordance with the view
that these flatfish are opportunistic feeders taking whatever is
available at the surface of the tidal flats (Kühl and Kuipers, 1979).
Virtually any species occurring at the Oosterschelde tidal flats
turns up in the food list of both species (Mandos, 1973). The second
conclusion is that plaice and flounder differ in their choice of
food. Note, for example, the differences between the occurrences of
the prey species Anaitides maculata, Nephtys hombergii, the tails of
Arenicola marina, Hydrobia ulvae, and Retusa obtusa.

Since it is unlikely that these opportunistic feeders are

Table 1. Numbers of Plaice Caught in Three Standing Nets Placed in Succession and Estimated Number of Plaice Meeting the First Net

	Attempt 1	Attempt 2	Attempt 3
Number caught in net 1.(a_1)	21	80	98
Number caught in net 2.(a_2)	25	64	98
Number caught in net 3.(a_3)	10	14	53
Estimated number meeting net 1.(N)	72	173	332

Table 2. Age Distribution of the Flatfish Caught on the Tidal Flats in the Period February 1972 - May 1973. The fish were caught every two weeks and the sample for age determination consisted of the total catch, or, if the catch was larger, of 25 individuals of each species.

Year-class	Plaice		Flounder	
	1972	1973	1972	1973
0	0	0	0	0
I	57	8	1	2
II	460	541	20	28
III	19	7	49	21
IV	0	0	44	8
V	0	0	13	4
VI	0	0	2	1
VII	0	0	1	0

specialised on different preys, a better explanation may be found in the differences in distribution of the macrobenthic species. Table 5 compares the abundance of seven benthic species at different tidal levels with the average number of individuals of these species in the stomach and guts of flatfish. In general, the differences in stomach contents of plaice and flounder match the differences in species composition of the benthos of low and high tidal flats respectively. In Table 5 Retusa obtusa is a notable exception, but

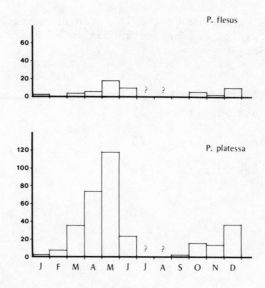

Figure 2. Average number of plaice and flounder caught per net per month. In July and August fishing was impossible because of enormous amounts of floating Ulva and other green algae.

Table 3. Average of Two-Weekly Catches and Densities of Flatfish on the Tidal Flats of the Grevelingen Estuary in March-May 1971 and of the Oosterschelde Estuary in March-May 1972 and 1973. Density calculations are explained in the text.

	Average number caught	Average number present	Average number per $10^4 m^2$
Plaice			
Grevelingen 1971	221	552	c. 276
Oosterschelde 1972	70	175	97
Oosterschelde 1973	111	277	154
Flounder			
Grevelingen 1971	26	65	c. 33
Oosterschelde 1972	7	17	9
Oosterschelde 1973	14	35	19

Table 4.   Food of Plaice and Flounder in the Oosterschelde Estuary
in 1972 and 1973. The first two columns indicate the
frequency of preys, expressed as the percentage of all
fishes, caught in two-weekly catches in the period February
1972 – May 1973, containing this prey. The last two
columns denote the average number of these prey items per
fish. Usually 25 fish were investigated from each two-
weekly catch, (unless the total catch was less than this
figure). Only the more important prey species are listed.

	Frequency (% of catch)		Number per fish	
	Plaice	Flounder	Plaice	Flounder
Anaitides maculata	56	0	3.1	0.0
Nereis diversicolor	29	31	0.4	1.0
Nephtys hombergii	50	8	0.8	0.1
Scoloplos armiger	33	29	3.9	4.8
Pygospio elegans	4	11	0.1	2.1
Arenicola marina – entire	19	13	0.2	0.2
Arenicola marina – tails	100	79	32.8	12.1
Lanice conchilega	7	11	0.8	0.1
Hydrobia ulvae	50	66	2.4	16.7
Retusa obtusa	51	0	3.4	0.0
Cerastoderma edule	71	58	3.5	6.3
Macoma balthica	45	50	1.0	2.4
Bathyporeia div. sp.	12	16	0.2	0.2
Gammarus sp.	27	19	0.4	0.6
Corophium sp.	61	48	1.9	2.6
Crangon crangon	15	21	0.3	0.3
Carcinus maenas	30	19	0.5	1.3

normally this species does not occur at high tidal levels.

Tentatively, the conclusion may be drawn that plaice and flounder feed at different levels of the tidal flats with the former species at the lower levels and the latter at the higher levels. This conclusion is supported by some limited data on flatfish caught at the same three tidal levels where the benthos was sampled (Table 4). At the highest level only flounder is caught whereas at the lower levels both species occur, as might be expected since both species have to migrate to the tidal channel at ebb. A distribution based on different types of sediment is unlikely in the Oosterschelde area investigated since the flats are fairly sandy from HW to LW level.

Kuipers (1977) found, for II-group plaice, a food-intake of about 4 mg ash-free dry weight per gramme fish (fresh weight) per 24 hours. From data by de Vlas (1979) slightly higher values may be inferred: 7 $mg.g^{-1}.24 h^{-1}$. For II-group plaice of about 100 g fresh weight this equals about 600 mg ash-free dry weight per fish per 24 hours. With the densities of plaice given in Table 3 this results in a predation of about 0.5-1.5 g ash-free dry $weight.m^{-2}$ in the period March-May. For the remainder of the year a figure of the same order of magnitude will hold. Compared to an estimated secondary production by the macrobenthos of over 50 g ash-free dry weight $.m^{-2}.year^{-1}$ for the Grevelingen estuary, of which the larger part is produced on the tidal flats (Wolff and de Wolf, 1977), it is evident that predation by II-group plaice, and by II-group and older flounder, on the benthos of the Delta estuaries was rather low in the period 1971-1973. This conclusion, however, rests heavily on our determination of net efficiency which may be wrong as has been pointed out above.

TIDAL MIGRATION AS A FEEDING STRATEGY

The catches of plaice and flounder on the tidal flats show clearly that these species use tidal migration as a feeding strategy. Figure 3 shows that in summer virtually all plaice leave the tidal channels during flood and return during ebb tide. In winter, however, plaice apparently stay in the tidal channels, which is in accordance with Figure 2.

During our investigations we caught plaice, flounder, and an occasional brill (Scophthalmus rhombus) on the tidal flats. Two other species of flatfish occurring abundantly in the Delta estuaries, were never caught on the flats, also not at night, viz. dab (Limanda limanda) and sole (Solea solea). Nevertheless, these species were caught in appreciable numbers along the edge of the tidal channel during fishing with a shrimp net for plaice. de Vlas (1979) records a few 0-group sole with empty stomachs and very few 0-group dab, brill, and turbot (Scophthalmus maximus) from the Balgzand tidal flats in the Wadden Sea. It appears that four common species of

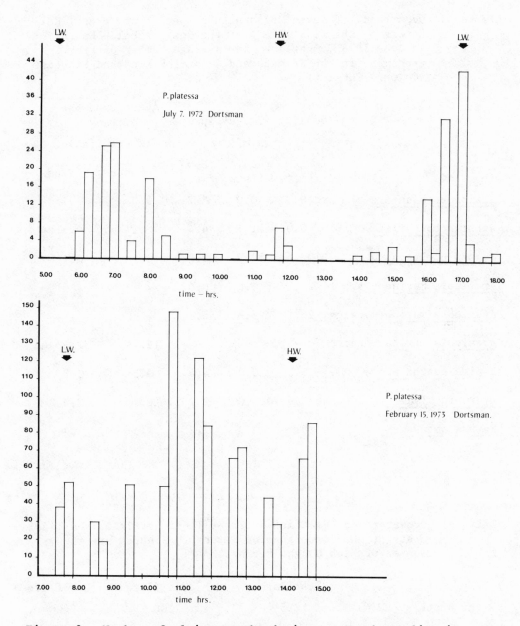

Figure 3. Number of plaice caught during repeated sampling in a tidal channel along the edge of the tidal flats. Top: samples from July 7, 1972. Plaice are mainly caught around low tide, which reflects tidal migration. Bottom: samples from February 15, 1973. Plaice are caught continuously, since there is no tidal migration.

Table 5.   Occurrence of Seven Macrobenthic Species at Three Tidal
           Levels on the Tidal Flats of the Oosterschelde Estuary
           in June 1973 Compared to the Average Number of Prey
           Specimens in the Stomach and Guts of Plaice and Flounder
           (Compare Table 4).

	Density of prey species $(\bar{x}.m^{-2})$			Average number of prey $(\bar{x}$ per fish)	
	High flats	Middle flats	Low flats	Flounder	Plaice
<u>Anaitides</u> <u>maculata</u>	0.0	0.0	6.7	0.0	3.1
<u>Nephtys</u> <u>hombergii</u>	0.0	11.1	4.4	0.1	0.8
<u>Pygospio</u> <u>elegans</u>	100.4	82.2	20.0	2.1	0.1
<u>Arenicola</u> <u>marina</u> - small	22.2	0.0	0.0	} 0.2	} 0.2
<u>Arenicola</u> <u>marina</u> - large	26.7	6.8	55.6		
<u>Arenicola</u> <u>marina</u> - tails	–	–	–	12.1	32.8
<u>Lanice</u> <u>conchilega</u>	0.0	0.0	4.4	0.1	0.8
<u>Hydrobia</u> <u>ulvae</u>	4575.6	7055.6	2.2	16.7	2.4
<u>Retusa</u> <u>obtusa</u>	8.9	0.0	2.2	0.0	3.4

Table 6.   Occurrence of Flatfish at Three Tidal Levels on the Tidal
           Flats of the Oosterschelde Estuary. Figures are the
           Number of Fish Caught Per Net.

	High flats	Middle flats	Low flats
Water depth at HW(m)	1.25	2.00	3.00
Sampling date	11 Aprl. 1973	7 June 1973	11 Aprl. 1973
Number of plaice	0	34	58
Number of flounder	3	5	3

flatfish use at least two different feeding strategies in the
polyhaline-euhaline parts of the Delta estuaries.

Strategy 1 is based on tidal migration. The animals feed only
on the tidal flats during high tide and during low tide they can be
found waiting along the edge of the tidal channel. Because of this
feeding migration they have to swim long distances (at least some
individuals swim 4 km per tide or more), but they are richly
rewarded by very frequent occurrence of potential prey on the tidal
flats. Failure in carrying out the migration properly almost certainly results in death because of desiccation, oxygen shortage or
bird predation. In Dutch and other NW-European estuaries plaice
and flounder can use this feeding strategy, but at other places,
e.g. in the Western Baltic (Arntz, 1980), the same species employ
the second strategy.

Strategy 2 is characterized by the absence of tidal migration.
These animals remain in subtidal areas and may feed continuously,
although light-dark rhythms may be involved as well. There is no
particular need to swim large distances, once a feeding area has
been discovered, but on the other hand food sources in the estuarine
subtidal habitat usually are sparse (Wolff and de Wolf, 1977). Dab
and sole are exhibitors of this strategy.

To some extent this use of different feeding strategies may be
considered as a way of niche partitioning. The visual hunters,
flounder, plaice and dab (de Groot, 1971) occupy in the Delta
estuaries three clearly different feeding niches, viz. the high
tidal flats, the low tidal flats and the subtidal area, respectively.
This situation offers many interesting ecological questions, for
instance on the energy balance of migrating and non-migrating species.
Even more interesting might be the situation in the more brackish
parts of the estuaries where only the flounder occurs. It seems
likely that here individuals of the same population use different
feeding strategies, viz. tidal migration and no migration. A study
of the energy balance of fishes in such a situation might be rewarding.

ACKNOWLEDGEMENTS

Many people helped with the field and laboratory work, but
especially Mr. W.J. Röber, at that time Skipper of the research-
vessel "Dr. J.G. de Man", spent many hours in support of this
project. Other important contributors were Messrs. P. van Boven,
P. de Koeyer and L. de Wolf. The late Dr. K.F. Vaas encouraged our
work and gave his valuable advice. Mr. C. Rappoldt gave advice on
the calculation of net efficiency. Dr. B. Kuipers and Mr. G.
Doornbos read the first draft of the manuscript.

REFERENCES

Arntz, W.E. 1980. Predation by demersal fish and its impact on the dynamics of macrobenthos. In: K.R. Tenore and B.C. Coull (eds.). Marine benthic dynamics. Belle W. Baruch Library Mar. Sci. 11: 121-149.

Beukema, J.J. 1976. Biomass and species richness of the macrobenthic animals living on the tidal flats of the Dutch Wadden Sea. Neth. J. Sea Res. 10: 236-261.

Bückmann, A. 1935. Ueber die Jungschollenbevölkerung der deutschen Wattenküste der Nordsee. Ber. dt. wiss. Komm. Meeresforsch. NF 7: 319-327.

Groot, S.J. de, 1971. On the interrelationships between morphology of the alimentary tract, food and feeding behaviour in flat-fishes (Pisces: Pleuronectiformes). Neth. J. Sea Res. 5: 121-196.

Kühl, H. and B.R. Kuipers, 1979. Qualitative food relationships of Wadden Sea fishes. In: N. Dankers, W.J. Wolff and J.J. Zijlstra (eds.). Fishes and fisheries of the Wadden Sea. Balkema, Rotterdam, p. 112-123.

Kuipers, B.R. 1973. On the tidal migration of young plaice (Pleuronectes platessa) in the Wadden Sea. Neth. J. Sea Res. 6: 376-388.

Kuipers, B.R. 1977. On the ecology of juvenile plaice on a tidal flat in the Wadden Sea. Neth. J. Sea Res. 11: 56-91.

Mandos, M.A. 1973. De invloed van de predatie door oudere schollen (Pleuronectes platessa) en botten (P.flesus) op de bodemfauna van zandig slik in het Deltagebied. Unpubl. rep. Delta Instituut v. Hydrobiologisch Onderzoek, Yerseke. 92 pp.

Nienhuis, P.H. 1970. The benthic algal communities of flats and salt marshes in the Grevelingen, a sea-arm in the south-western Netherlands. Neth. J. Sea Res. 5: 20-49.

Swennen, C. 1976. Waddenseas are rare, hospitable and productive. In: M. Smart (ed.). Proc. Int. Conf. Conservation Wetlands Waterfowl, Heiligenhafen, 1974. International Wildfowl Research Bureau, Slimbridge, p. 184-198.

Vlas, J. de 1979. Annual food intake by plaice and flounder in a tidal flat area in the Dutch Wadden Sea, with special reference to consumption of regenerating parts of macrobenthic prey. Neth. J. Sea Res. 13: 117-153.

Wells, B., D.H. Steele and A.V. Tyler, 1973. Intertidal feeding of winter flounders (Pseudopleuronectes americanus) in the Bay of Fundy. J. Fish. Res. Board Canada 30: 1374-1378.

Wimpenny, R.S. 1953. The plaice, being the Buckland lectures for 1949. Arnold, London. 145 pp.

Wolff, W.J. 1973. The estuary as a habitat. An analysis of data on the soft-bottom macrofauna of the estuarine area of the Rivers Rhine, Meuse and Scheldt. Zoologische Verhandelingen 126: 1-242.

Wolff, W.J. and L. de Wolf, 1977. Biomass and production of zoo-

benthos in the Grevelingen estuary, the Netherlands. <u>Estuar</u>.
<u>Coast</u>. <u>Mar</u>. <u>Sci</u>. 5: 1-24.

ON CROPPING AND BEING CROPPED: THE REGENERATION OF BODY PARTS BY

BENTHIC ORGANISMS

J. de Vlas

Netherlands Institute for Sea Research, Texel
Present address: Research Institute for Nature Management, P.O. Box 59, Texel, Netherlands

ON CROPPING

     To get food or oxygen, many benthic organisms expose body parts
like siphons and tentacles near to, or above the sediment level.
Others, like the deep-living worm Arenicola marina, defaecate at the
surface of the sediment, thus exposing their tail tips. As a consequence of their opportunistic feeding strategy, cropping of such body
parts can yield an important food source for flounder (Platichthys
flesus) and juvenile plaice (Pleuronectes platessa) (de Vlas, 1979a).
In the westernmost part of the Wadden Sea, called Balgzand, about one
third of the total food intake of these fish consisted of siphon tips,
tentacles, tail ends and heads of molluscs and polychaetes. Figure
1 shows a number of organisms from which body parts (arrows) were
found in flatfish stomachs. This paper briefly considers this cropping from the points of view of the food source they represent, secondary production and as losses which the invertebrates have to cope
with.

     Since benthic organisms easily regenerate such body parts, they
can produce biomass which is not reflected in increase of body weight
or in absence of these body parts. In terms of food chain transfer
from benthic animals to a next trophic level, these cropped parts can
play an important role. This is illustrated in Table 1, in which the
biomass involved in flatfish cropping is compared with food chain
transfers like predation by birds, or elimination by micro-organisms.
These can be traced because they are connected with mortality in the
benthic populations. Flatfish cropped mainly body parts of Arenicola
marina, Macoma balthica, Cerastoderma edule and Heteromastus
filiformis. These estimates of regenerating body parts concern the
consumption by flatfish only, and so they are minimum values. For

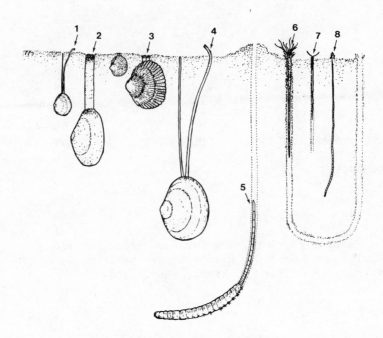

Figure 1. Organisms in the investigated area, from which body parts
were found in stomachs of plaice and flounder. Arrows indicate the
cropped parts of 1. Macoma balthica; 2. Mya arenaria; 3. Cerastoderma edule; 4. Scrobicularia plana; 5. Arenicola marina; 6.
Lanice conchilega; 7. Pygospio elegans and other small spionid polychaetes; 8. Heteromastus filiformis.

instance, birds as well as the worm Nereis diversicolor, can prey on
tail tips of Arenicola (Barnes, pers. comm.; Witte and de Wilde,
1979) and siphon tips of Macoma have also been found in stomachs of
Pomatoschistus minutus (Verbeek, 1976).

ON BEING CROPPED

     On the benthic side of the food chain, the effects of loss of
siphon tips and tail tips were studied for Macoma balthica and
Arenicola marina. Both organisms proved to be well adapted to flatfish cropping.

     Repeated artificial amputation of Macoma siphon tips led to a
rapid regeneration (Figure 2), which was faster than the calculated
flatfish siphon tip consumption per Macoma in the tidal flat area.
This agrees with the increasing siphon weights on Balgzand during
the 3-month period (April to June) when flatfish prey on siphon
tips.

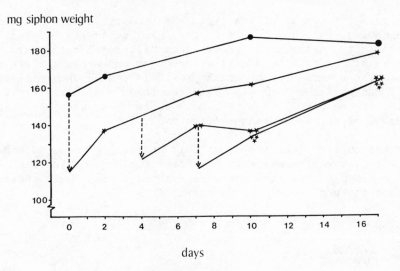

Figure 2. Mean changes in siphon weight of siphons of <u>Macoma balthica</u>
amputated once, twice and thrice in a laboratory experiment. ● =
controls, ∗ once; ∗∗ twice and ∗∗∗ thrice amputated siphons.

Table 1. Approximate Annual Elimination of Biomass by Mortality
          (= complete individuals) and by Flatfish Cropping (=
          regenerating parts of individuals) in 1976 on the East
          Side of the Balgzand Tidal Flat Area.
          Weights in grammes ash-free dry wt m^{-2}. Benthos data from
          J.J. Beukema.

	mortality	cropped
<u>Macoma</u> <u>balthica</u>	1	0.1
<u>Cerastoderma</u> <u>edule</u>	12	0.5
<u>Mya</u> <u>arenaria</u>	3	0.1
<u>Arenicola</u> <u>marina</u>	4	1.3
<u>Heteromastus</u> <u>filiformis</u>	1	0.03
Other macrobenthos	9	0.1

     For <u>Arenicola</u>, the tail tip losses did not prevent growth of
the tails, but the increase in tail weights lagged behind when compared to the increase in trunk weights (de Vlas, 1979b). Tail tip
losses could be detected by the decreasing numbers of tail segments

per worm in the course of its life. Young worms mostly had about
100 segments; old ones (about 5 years old) generally had only a few
segments left. Experiments showed that tail regeneration (by growth
of the remaining tail segments) occurred at the expense of trunk
growth. The retardation of trunk growth increased when a larger
part of the tail was removed. This resulted in a fairly rapid res-
toration of the normal body proportions, in which the tail weight
was 20-25% of the trunk weight (Figure 3).

These observations show that this cropping of exposed body parts
of benthic invertebrates is an important food source for fish and
that the cropped organisms are well adapted to deal with these losses
by possessing good powers of regeneration.

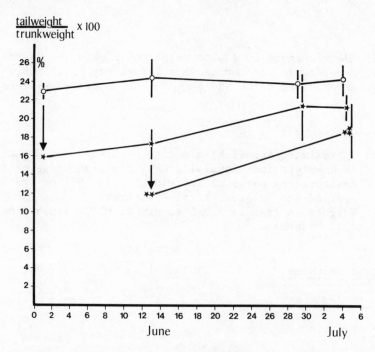

Figure 3. Mean change in the ratio between tail and trunk weight
after artificial amputation of lugworm tails. Arrows indicate the
effect of amputation. o = controls; * once amputated and ** twice
amputated groups of 30 worms each, protected by a cage to prevent
flatfish predation. Bars denote standard errors.

REFERENCES

Verbeek, F.A. 1976. Aantallen, groei, productie en voedselopname
    van de zandgrondel (P.minutus) en de wadgrondel (P.microps) op

het Balgzand. Unpubl. rep. Netherlands Institute for Sea
     Research, 1976-9.
Vlas, J. de, 1979a. Annual food intake by plaice and flounder in a
     tidal flat area in the Dutch Wadden Sea, with special reference
     to consumption of regnerating parts of macrobenthic prey. Neth.
     J. Sea Res. 13(1): 117-153.
Vlas, J. de, 1979b. Secondary production by tail regeneration in a
     tidal flat population of lugworms (Arenicola marina), cropped
     by flatfish. Neth. J. Sea Res. 13(3/4): 362-393.
Witte, F. and de Wilde, P.A.W.J. 1979. On the ecological relation
     between Nereis diversicolor and juvenile Arenicola marina.
     Neth. J. Sea Res. 13(3/4): 394-405.

HOW FORAGING PLOVERS COPE WITH ENVIRONMENTAL EFFECTS ON INVERTEBRATE BEHAVIOUR AND AVAILABILITY

M.W. Pienkowski

Department of Zoology
University of Durham
South Road, Durham, DH1 3LE

INTRODUCTION

Estuarine sand and mudflats hold dense populations of invertebrate animals which are preyed upon by fish while the tide is high and by birds during the low water period. Their principal predators are shore birds which face the problems that the invertebrates may be hidden under a layer of mud or be buried out of reach of their bills at some depth in the soft sediment.

The main group of shorebirds, the waders (Charadrii), have fine bills which are used to pick up individual prey animals. Two basic strategies (with many gradations between these) are used by waders to detect and catch their prey. One group, including many of the sandpiper family (Scolopacidae) and the Oystercatcher Haematopus ostralegus, commonly include a tactile element in their foraging: the bill is placed in contact with the substratum at various locations as the bird walks over the area. The birds may also use sight, sometimes choosing where to place the bill in response to visible cues. The other group of waders, typified by the plovers (Charadriidae), use a totally visual method of foraging (Burton, 1974; Pienkowski,1980b). These normally stand still for a time while scanning the area for indications of available prey, and may then run rapidly to peck at a prey item,or, if none is seen, run to a new waiting position (Fig. 1). Tactile foragers usually have long bills with high densities of pressure detectors (Herbst corpuscles) along much of their lengths to aid probing into the substratum; plovers, and other visual searchers, have relatively short bills with concentrations of Herbst corpuscles mainly at the tip, this combination being compatible with rapid alignment of the head and bill towards the prey, hard impacting on the substratum

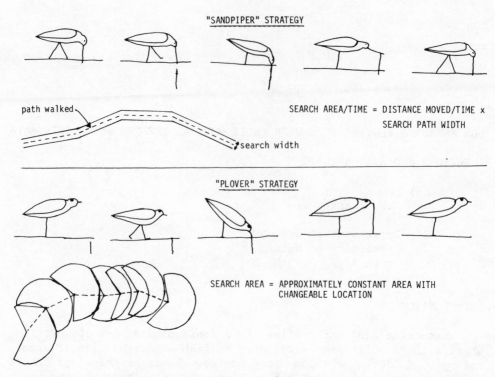

Figure 1. Generalized foraging patterns of tactile-searching sand-piper and visual-hunting plover.

surface and precise gripping of prey (Bolze,1968; Burton,1974; Pienkowski,1980b).

These two basic strategies are associated with different bene-fits and disadvantages, as I have discussed elsewhere (Pienkowski, 1980a,b). Generally, the 'plover strategy' depends on high activity of prey, particularly that which causes it to come frequently to the surface, while the 'sandpiper strategy' depends on prey being near enough to the surface to be within range of the bill. As the activity of prey is generally more sensitive to temperature variations than is the depth distribution, the 'sandpiper strategy' is superior to the 'plover strategy' at low temperatures. (In mid-winter, plovers in NE England have to forage for most of the time that the sand-flats are exposed by the tide in order to meet their energy requirements and, indeed, on some days it is likely that this is impossible – Pienkowski,1980b; Dugan, this volume). However, because the range of detection and potential capture by the 'plover strategy' is much greater than when foraging tactilely, this has advantages at low prey densities. From an analysis of prey taken in different areas,

it is clear that although individual wader species may have preferences for certain prey (e.g. Goss-Custard,1977), they are not specialized to take only particular prey species (in contrast to the frequently quoted and illustrated supposed relationship between bill length
and depth of prey animal, e.g. Green,1968). They differ rather in
the way they exploit the behaviours of their prey (Pienkowski,1980a,b).

In this paper, I shall summarize the ways in which environmental
conditions, particularly temperature, affect the behaviour of some
intertidal invertebrates and how plovers modify their foraging behaviour in response to this (including one way in which they may increase
the availability of the prey animals by influencing their behaviour).
These considerations of prey availability have relevance also to
other shorebirds but, for the reasons given above, plovers are more
susceptible to decreases in prey activity.

METHODS

Field observations of the behaviour of shorebirds and their prey
were made on the Lindisfarne National Nature Reserve, Northumberland
(55°40'N, 1°50'W) in the years 1973-76 by myself, except for those
on the lugworm Arenicola marina which were made in the same area by
P.C. Smith in 1970-73 (Smith,1975). He recorded the activity of
Arenicola by noting the frequency of defaecation (cast productions);
I measured that of the capitellid polychaete Notomastus latericeus
by counting the frequency of outflows of water from small holes
(diameter < 1 mm) in the surface of the mud. These outflows occur
when the worm moves to the surface (Pienkowski,1980b). I also
measured the activity levels of small Crustacea, particularly the
amphipod Bathyporeia and the isopod Eurydice by the numbers of these
animals moving per unit time in unit area (the animals being un-
detectable when still and hidden in the surface layer of sand).
Further details of methods are given by Smith (1975) and Pienkowski
(1980b), who also argue that these signs are the cues used by
visually-hunting shorebirds to detect the prey. Sand temperatures
were measured at 2 cm depth. Although temperatures may vary with
depth, those at different depths were correlated, and temperatures
measured at 2 cm provide a reliable index of the temperatures
experienced by the worms.

Plovers were watched from a car parked on the sand flats and
moved when necessary. Their behaviour, pecks and prey taken were
noted on a continually-running tape recorder or on ciné-film, and
the sequence of activities and durations of each activity later
noted or calculated and transferred to computer storage. The
procedure and checks on validity and reliability are detailed by
Pienkowski (1980b), who found that well over 90% of pecks resulted
in a prey item being taken.

ENVIRONMENTAL EFFECTS ON THE ACTIVITY OF INTERTIDAL INFAUNA AND THE AVAILABILITY OF THESE TO SHOREBIRD PREDATORS

For most of the time of tidal exposure many polychaetes are buried too deep in the substratum to be reached by birds' bills. However, some worms make periodic excursions towards the surface. For example, Arenicola may move to the surface of the tail shaft of its burrow to defaecate. Notomastus also moves to the surface in its vertical burrows for as yet unknown reasons. Very infrequently their heads emerge from the holes and they feed but normally they do not become visible at the surface. For both polychaetes, the periods spent at the surface are brief relative to the time at depth. It is not known whether this is due to physiological limit- ations on the worms when the water table is low or to avoidance of predation by shorebirds or both.

Some other worms, including the polychaete Phyllodoce maculata on the sand flats at Lindisfarne, may partly emerge to feed at the sand surface for longer periods, especially in water filled hollows. However, this behaviour is far more common at night than by day, presumably to avoid predation. The nocturnal behaviour of intertidal invertebrates and foraging plovers is considered further by Dugan (this volume) and Pienkowski (1980b); the present paper deals mainly with the situation in daylight.

The surface activity in the field of all sand flat invertebrates studied so far at Lindisfarne is highly temperature dependant (Fig. 2), generally increasing as temperatures rise over the range -2 to 17°C (the limits of observation). This is true of surface living Crustacea as well as burrowing worms.

The activity of many invertebrates also increases with the wet- ness of the substratum. This gives rise to a tidal activity pattern in Arenicola, with (at any one moment) the highest proportions at the surface near the tide edge, wherever this might be (Smith,1975). Any tidal pattern in surface signs of Notomastus activity is less marked. Although the proportions at the surface at any one moment may be slightly higher when the water table is high, the outflows of water from the holes are difficult to see (probably for birds as well as for humans) when the sand surface is covered by a film of water (Pienkowski,1980b).

Increasing wind strength tends to cause more rapid drying of the substratum, and hence to reduce surface activity of invertebrates away from the tide edge. Wind may also reduce visibility of cast formation by Arenicola in shallow water, because of increased wave action (Smith,1975; Evans,1976).

Rainfall may also decrease detectability of prey, both by decreasing the range to which activity is visible and because the

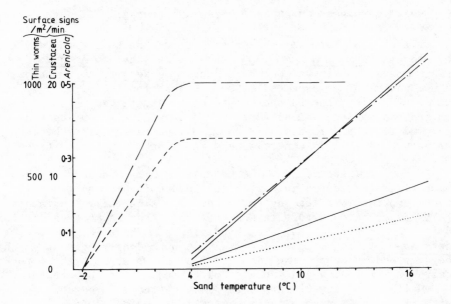

Figure 2. Rates of appearances of surface signs of various invertebrates at Lindisfarne in relation to temperature.
Arenicola during ebbing (——— ———) and flowing (— — — — —) tides
(Smith,1975); ——————— Notomastus at two sites (regression equations
are y = 84.8x - 291, P < 0.001 and y = 33.5x - 110, P < 0.001
(Pienkowski,1980b); and small Crustacea, without vibration ········
(y = 0.43x - 1.4, P < 0.001) and with vibration (see text) —•—•—
(y = 1.62x - 5.0, P < 0.001) (Pienkowski,1980b).

impacting raindrops are difficult (at least to human eyes) to distinguish from outflows from holes of Notomastus. For the latter
reason, it is difficult to investigate whether rainfall also has an
effect on Notomastus surface activity.

     Generally then, decreasing temperatures, increasing windforce,
increasing rainfall, and passage of time after tidal exposure tend
to reduce the availability or detectability of intertidal prey to
shorebirds. The first two factors also increase the metabolic energy
demand of shorebirds, so exacerbating the situation (see also Dugan
et al.in press; Davidson, this volume). In high winds, the plovers
may also suffer difficulties in making the rapid directed movements
which are crucial to their foraging strategy. In the following
sections, I shall explore the ways in which the feeding rates of
plovers are affected by such conditions and how the plovers modify
their behaviour to exploit the available prey.

ENVIRONMENTAL EFFECTS ON FEEDING RATES OF PLOVERS

    On the sand flats at Lindisfarne during the autumn and winter,
Grey Plovers Pluvialis squatarola, take relatively few Arenicola and
large numbers of thin red worms, principally Notomastus (Pienkowski,
1980b). As might be expected from considerations of prey behaviour
and detectability, the rate of capture of thin red worms increased
with increasing sand temperature, but decreased with an increase in
both wind force and rainfall (Figs. 3-5). This supports the idea
that prey availability controls feeding rates in such conditions.

    In spring (March to May), Arenicola is more frequently taken
than earlier in the season. This may be due in part to generally
higher temperatures, leading to increased availability, but also
results from the departure in February and March of most Bar-tailed
Godwits, Limosa lapponica, from Lindisfarne. These long-billed
waders are able to forage effectively in high-density groups at the
tide edge and, thereby, exclude Grey Plovers whose foraging method
requires a large area available to each bird. After departure of
the Godwits, the Grey Plovers spread nearer the tide edge where
availability of Arenicola is highest. In these conditions, their
rate of capture of Arenicola was directly correlated with sand
temperature (r = 0.30, P < 0.001) and negatively correlated with
time after tidal exposure (r = -0.30, P < 0.01), again matching the
activity patterns of the prey.

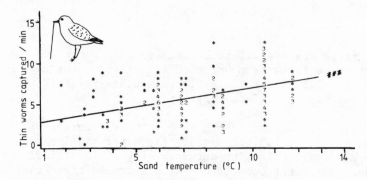

Figure 3. Estimated rate of capture of thin worms by Grey Plovers
on the low flats at Lindisfarne in autumn and winter, in relation to
sand temperature. Fitted regression is y = 0.36x + 2.49 (P < 0.001).
(On this and similar plots, asterisks indicate single points and
numerals the number of coincident points; '9' indicates 9 or more
such points).

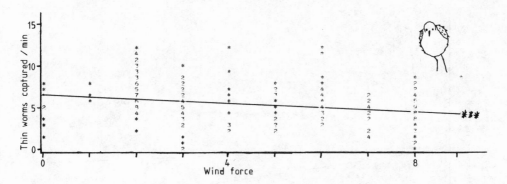

Figure 4. Estimated rate of capture of thin worms by Grey Plovers
on the low flats at Lindisfarne in autumn and winter, in relation
to wind force (Beaufort scale). Fitted regression is y = 6.63 -
0.26x (P < 0.001).

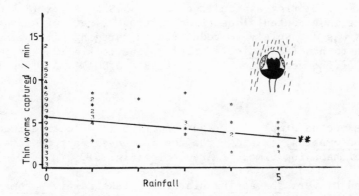

Figure 5. Estimated rate of capture of thin worms by Grey Plovers
on the low flats at Lindisfarne in autumn and winter, in relation
to rain-fall. (Rainfall assessed during observations on a subjective
scale in which 0 = no rain; 3 = light rain; 5 = heavy rain).
Fitted regression is y = 5.63 - 0.44x (P < 0.01).

    In the same period of the year, however, the rates of capture
of Notomastus decreased with increasing temperature (r = -0.37,
P < 0.001), in opposition to the increased prey activity with
temperature, and decreased with time after high water (r = 0.48,
P < 0.001). This would be expected, however, if the Plovers were
selecting Arenicola, which are very large, in preference to the
much smaller Notomastus. The rates of capture of the two types of
prey were, indeed, negatively correlated (Fig. 6). Furthermore, the
rate of capture of Notomastus was suppressed significantly more than

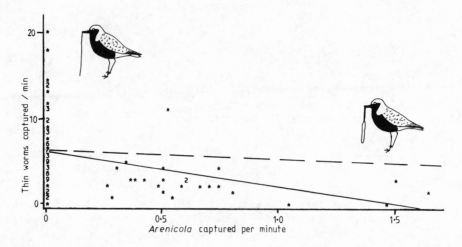

Figure 6. Estimated rate of capture of thin worms by Grey Plovers
in spring on the low flats in relation to the rate of capture of
Arenicola. Fitted regression line is y = 5.93 - 4.12x (P < 0.001).
Dashed line is expected slope if depression of rate of capture of
thin worms was due solely to reduction in foraging time because of
time taken to handle Arenicola. This slope is significantly different
from that of the regression (t = 3.71, P < 0.001).

would be expected simply from the reduction in effective foraging
time for Notomastus, arising from the increased capture rate and
much longer handling times of Arenicola (Fig. 6).

    Ringed Plovers Charadrius hiaticula, take mainly Notomastus
and other similar worms on their main low water feeding area. At
temperatures below 6°C, the rate of capture of such worms decreased
with decreasing temperature (Fig. 7) and increasing windforce
(r = -0.19, P < 0.01), as expected from the behaviour of the prey
if the availability of such prey were influencing feeding rate.
However, at higher temperatures, these relationships were reversed,
capture rate decreasing with increasing temperature (Fig. 7) and
decreasing wind force (r = 0.23, P < 0.01). This appears again to
be associated with increasing selectivity for larger worms as
temperature increases (Fig. 8), and led to maintenance (rather than
any decrease) of the rate of energy intake with increasing temperatures (Pienkowski,1980b). The idea that Ringed Plovers were more
selective about 6°C receives some support from a comparison between
the rate of capture of worms and the rate at which their outflows
appeared in the search area (see Pienkowski,1980b). At lower
temperatures the birds must have pecked at almost all outflows,
whereas at higher temperatures some outflows must have been
ignored (Fig. 7).

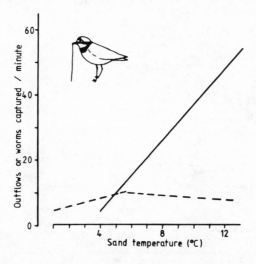

Figure 7. Estimated numbers of outflows from thin worm holes in
search area of Ringed Plover (solid line) and estimated rate of
capture of thin worms by Ringed Plovers (dashed lines), in relation
to sand temperature on the low flats at Lindisfarne. Solid line
based on the same data as Fig. 2; dashed lines on observations of
foraging birds. Regression equations of dashed lines are
$y = 1.19x + 3.65$ (P < 0.001) below 6°C, and $y = 11.03 - 0.27x$,
(P < 0.001) above 6°C.

MODIFICATIONS OF FORAGING BEHAVIOUR

The time spent by plovers in the pause position before moving
to take prey (the 'waiting time') tended to be longer for small prey
than for large prey, and the 'giving-up' time (i.e. the time before
the plover moves to a new standing position without taking prey)
longest of all (Fig. 9). This relationship, which applied to both
plover species and in different parts of the sand flats, is at first
sight surprising as cues indicating the availability of small prey
occur more frequently than those of larger prey (mainly because
smaller prey occur at higher densities in any particular area). If
plovers always took the first prey then small prey would predominate
and no selection would result. However, if their readiness to react
to cues from smaller prey increased progressively from an initially
very low level during each waiting period, birds would increase
their chances of taking large prey, if present in the area searched.
Even if large prey are not available, some intake is likely and
the behaviour will result in many areas being sampled. This readiness to move could vary with the density of available prey, as is
suggested by a high correlation between waiting and giving-up times
for both plover species in each study area (Pienkowski,1980b).

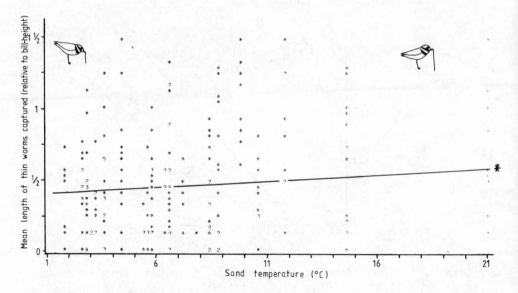

Figure 8. Mean lengths (relative to height of bill off the ground)
of thin worms taken by Ringed Plovers on the low flats at Lindisfarne in relation to sand temperature. Height of bill off the ground
(in 'normal' stance) is about 8 cm.

Indeed, the waiting and giving-up times of Ringed Plovers tended
to decrease as temperatures rose to about 6°C. For Grey Plovers,
mean waiting times increased as the rate of taking Arenicola
increased.

The distances moved to a new waiting position also changed
according to feeding conditions (Fig. 10). Generally, the distance
moved after taking a prey item was inversely related to the energetic
value of that prey, and was greatest when moving to a new waiting
position after no prey was taken. The latter distance tends to be
just far enough to scan a new search area (i.e. they move just
further than the normal maximum range of running to take prey,
Pienkowski,1980b). Such a strategy would be adaptive in keeping
the birds in an area of high density of available prey if prey
distributions are clumped, as indeed they tend to be.

In some situations of low prey availability, plovers may be
able to increase the activity of their prey, thus making them more
detectable. This appears to be the reason for foot-vibration in
which the bird stands on one foot and rapidly vibrates the other
with toes just in contact with the ground. At Lindisfarne, the
plovers use this behaviour mainly on the flats at high tidal levels.
Experimental simulation of this behaviour at these sites led to
increased activities of small Crustacea (Fig. 2). This probably

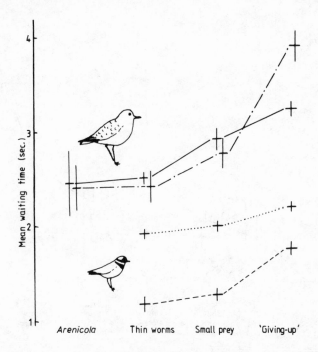

Figure 9. Mean (± 1 s.e.) waiting times in the pause position before
taking various prey, or 'giving-up' and moving to a new pause
position. ─·─·─, ───────── Grey Plover on the higher, and
lower parts of the flats; ······· ,─────── Ringed Plovers on the
higher, and lower parts of the flats.

results from the swimming reactions of prey in response to vibrations
normally associated with the approaching tide edge as well as in
response to the liquefying of the thixotropic sand (see Sparks,1961;
Enright,1962, 1965; Jones & Naylor,1970; Pienkowski,1980b). Although the average pecking rates of Ringed Plovers using and not
using foot-vibration did not differ when observations under all
conditions are combined (Pienkowski,1980b), foot-vibration tended
to occur in conditions associated with low prey availability - low
temperature, high winds or rain (Table 1). Thus, the birds probably
used foot-vibration to increase prey activity (perhaps to an
acceptable or optimal level) when this would otherwise be reduced.

PROGRAMME FOR A PERFECT PLOVER: A STEP-BY-STEP GUIDE

     The strategy which these studies suggest is used by foraging
plovers may be summarized as a set of 'rules':

1.   Plovers forage by scanning a search area from a standing
      position and running rapidly to peck at prey detected.

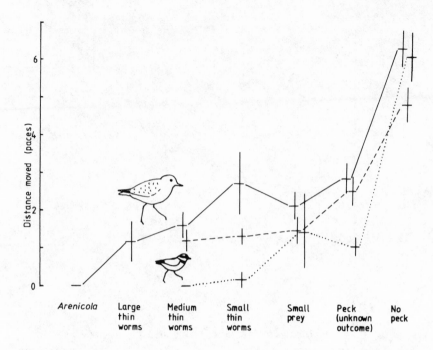

Figure 10. Mean (± 1 s.e.) distances moved to new waiting position
in relation to the outcome of the preceding peck (or no peck if the
plover 'gave-up' at the last site). ————— Grey Plovers on the
low flats (mean pace length about 14 cm); — — —, •••••Ringed Plovers
on the low, and high flats (mean pace length about 10 cm).

2.   In adverse conditions (low temperatures, high winds, rain, dry
      substratum) the feeding rate of plovers is limited by the rate
      of appearance of cues resulting from the surface activity of
      suitable prey.

3.   In some situations (e.g. Ringed Plovers feeding on small
      Crustacea) the plovers may increase the activity of prey by
      foot-vibration.

4.   As conditions improve leading to increases in prey activity,
      and hence availability, plovers become more selective for
      larger prey (at Lindisfarne, large Notomastus for Ringed
      Plover and also Arenicola for Grey Plovers).

5.   Selection is achieved by becoming progressively more ready,
      within each waiting period, to take a small prey item or
      eventually to move to a new waiting site without taking a
      prey item.

6.   The feeding rates and waiting times appear to be monitored

Table 1   Mean Environmental Conditions During Observation of Foot-
         Vibration Behaviour

	Birds foot-vibrating for some or all time		Birds not foot-vibrating
<u>North Shore</u>			
Air temperature	10.4 ± 0.8 (32)	***	14.7 ± 0.4 (141)
<u>High Flats near Rig</u>			
Air temperature	11.2 ± 1.1 (20)	ns	15.5 ± 4.9 (4)
<u>High Flats near small</u> <u>salt marsh</u>			
Air temperature	5.3 ± 0.2 (34)	ns	5.7 ± 0.7 (40)
Wind force	4.8 ± 0.5 (34)	*	3.7 ± 0.3 (52)
Rain fall	0.9 ± 0.3 (34)	*	0.2 ± 0.1 (52)

rapidly to adjust giving-up times to local conditions.

7. Sampling of a variety of areas but concentration at profitable
    sites is achieved by adjusting the distance moved after taking
    prey inversely to the value of that prey.

8. If no prey is taken, the plover moves far enough to scan a
    new search area.

9. The plovers avoid concentrations of other birds, particularly
    tactile-foraging sandpipers, as visual foraging requires a
    large search area per bird.

These rules are explored more fully and extended to consider
flock structure, distribution and nocturnal feeding elsewhere
(Pienkowski 1980b).

ACKNOWLEDGEMENTS

I am grateful to: the Research and Special Publications Fund
of the British Ornithologists' Union for financial support; the
Nature Conservancy Council for permission to conduct studies at
Lindisfarne National Nature Reserve; Dr. P.R. Evans for valuable
advice and encouragement throughout the project and criticisms of
a draft of this paper; my colleagues N.C. Davidson, P.J. Dugan,
P.C. Smith and D.J. Townshend for information and discussion on

various points; my wife, Ann, for help at all stages of the study.

REFERENCES

Bolze, G. 1968. Nordung und Bar der Herbstschen Körperchen in
    Limicolenschnäbeln im Zusammenhang mit der Nahrungsfindung.
    Zoologischer Anzeiger 181: 311-355.
Burton, P.J.K. 1974. Feeding and the Feeding Apparatus in Waders.
    Brit. Mus. (Nat. Hist.), London.
Dugan, P.J., Evans, P.R., Goodyer, L.R. and Davidson, N.C. in press.
    Winter fat reserves in shorebirds: distance of regulated levels
    by severe weather conditions. Ibis.
Enright, J.T. 1962. Responses of an amphipod to pressure changes.
    Comp. Biochem. Physiol. 7: 131-145.
Enright, J.T. 1965. Entrainment of a tidal rhythm. Science 147:
    864-867.
Evans, P.R. 1976. Energy balance and optimal foraging strategies
    in shorebirds: some implications for their distributions and
    movements in the non-breeding season. Ardea 64: 117-139.
Goss-Custard, J.D. 1977. The energetics of prey selection by red-
    shank, Tringa totanus (L.), in relation to prey density.
    J. Anim. Ecol. 46: 1-19.
Green, J. 1968. The Biology of Estuarine Animals. Sidgewick &
    Jackson, London.
Jones, D.A. & Naylor, E. 1970. The swimming rhythm of the sand-
    beach isopod Eurydice pulchra. J. exp. mar. Biol. Ecol. 4:
    188-199.
Pienkowski, M.W. 1980a. Differences in habitat requirements and
    distribution patterns of plovers and sandpipers as investigated
    by studies of feeding behaviour. Proc. IWRB Feeding Ecology
    Symp., Gwatt, Switzerland, Sept. 1977. Verhn. Orn. Ges. Bayern
    23: in press.
Pienkowski, 1980b. Aspects of the Ecology and Behaviour of Ringed
    and Grey Plovers Charadrius hiaticula and Pluvialis squatarola.
    Ph.D. Thesis, University of Durham.
Smith, P.C. 1975. A study of the winter feeding ecology and behaviour
    of the Bar-tailed Godwit (Limosa lapponica). Ph.D. Thesis,
    University of Durham.
Sparks, J.H. 1961. The relationship between foot-movements and
    feeding in shorebirds. Brit. Birds 54: 337-340.

PREY DEPLETION AND THE REGULATION OF PREDATOR DENSITY: OYSTERCATCHERS

(HAEMATOPUS OSTRALEGUS) FEEDING ON MUSSELS (MYTILUS EDULIS)

L. Zwarts and R.H. Drent*

Rijksdienst voor de Ijsselmeerpolders, Postbox 600,
8200 AP Lelystad, Netherlands
*Zoological Laboratory, University of Groningen,
9750 NN Haren, Netherlands

INTRODUCTION

One of the many decisions facing birds attempting to maximize their food intake is the choice of where to go to feed. Certainly they should restrict their search to where food is plentiful, and concentration in the optimal parts of the feeding area has been documented for the Oystercatcher by Hulscher (1976) and Goss-Custard (1977). On the other hand, birds should avoid high feeding densities as the presence of conspecifics may in itself depress the feeding rate, a direct and immediate effect not dependent on prey depletion. This inhibitory effect has previously been demonstrated in other birds feeding on intertidal areas (Redshank, Tringa totanus, Goss-Custard, 1976; Curlew, Numenius arquata, Zwarts, in press). The outcome of these opposing tendencies expresses itself in a compromise, whereby the richest feeding areas are filled first, and as increasing numbers of birds utilize the area, marginal feeding areas come into use. This "buffer effect" is a familiar concept in discussions on the function of territoriality (Kluijver and Tinbergen, 1953; Fretwell and Lucas, 1970) and appears to play a role in the spacing of birds over the intertidal feeding area (Zwarts, 1976; Goss-Custard, 1977).

This paper concerns Oystercatchers feeding on intertidal mussel banks, and deals with the following questions: (1) is the presence of other birds a factor depressing intake rate? (2) can a "buffer effect" be demonstrated when mussel banks at various distances from the roost are compared? (3) do the birds feed in a higher density on the mussels if the food supply is more dispersed? (4) do the birds in their year-round utilization of the mussel bank have a depleting effect on the prey population? We have emphasized the

effect bird density exerts, as we are interested in the implication
that measurable effects on intake rate might provide a proximal
mechanism important in achieving the adjustment of bird density over
the available habitat units.

METHODS

The study was performed on two intertidal mussel banks south
of the Frisian island of Schiermonnikoog. Between 1971 and 1973
thirteen plots of 1 ha and 14 plots of 0.33 ha were pegged out, and
with the aid of students participating in an annual field course,
observations on one of these plots have continued in most years since
then. Within the plots the coverage by mussels was determined by
laying down a grid of ropes and measuring the linear distance where
mussels were present (coverage in the plots varied from 1%, bare
mudflats with scattered mussel clumps, to 50%, massive mussel beds
interspersed with very soft mud, small creeks and tidal pools).
Between May and November 1973 (and in certain periods annually since
then), the benthic fauna was sampled each month, samples (15-40)
being taken from the mussel beds using a core sampler ($1/64$ m^2 area).
For mussels larger than 40 mm and hence large enough to form the
prey of the Oystercatcher, the standard error of the mean density
turned out to be 8-15% relative to the mean obtained from this
sampling procedure, i.e. the method is sensitive to consistent shifts
in mussel density. For converting intake figures into weight of food
consumed, regressions for log weight in relation to log shell length
were determined for mussel samples pooled per month (hence a mussel
of a given size was assumed to have the same weight of flesh content
for all plots). Ash-free dry weights (AFDW) were determined following
standard laboratory procedures.

Birds on the mussel banks and neighbouring mudflats were observed
from hides erected on towers approximately 4 m in height. Observers
entered the tower by wading out as the tide began to recede, and
counted the birds in the plots every quarter of an hour until the
tide came in again and drove the birds away. Capture rate was measured in between the counts by observing the time (in seconds) to
search and eat three mussels, or by observing the number of mussels
taken by actively feeding birds followed for ten minutes (feeding
is here defined as total time minus preening and resting, and hence
includes aggressive encounters). The size of mussel taken was determined by collecting fresh shells recently opened (as revealed by
elasticity of the hinge), and feeding rates are expressed as g AFDW
per time unit (using the weight/length relationship).

In this particular area, human impact on the mussels was nil,
since there was no fishing on mussels.

Size Selection

When mussels taken by Oystercatchers are compared with the size distribution of the mussels on offer (Figure 1) a clear preference for the larger mussels appears: when compared to the "modal mussel" of 44 mm in 1973, the relative risk of a mussel of 50, 54 or 58 mm being taken by Oystercatchers was 3.6, 6.7 and 10.5 times as high respectively. Virtually all Oystercatchers at Schiermonnikoog opened mussels by jabbing the bill between the shell valves and had thus to find mussels in shallow water gaping very slightly. Only rarely (less than 1% of shells collected) were shells opened by hammering with the bill. That selection of large mussels is indeed profitable, especially when the stabbing method is used, is shown in Figure 2A. Although search time increases when hunting selec- tively, the yield per time unit expended on prising the mussel and extracting the flesh from the shell ("handling", Figure 2B) shows a steep rise with mussel size. Since on average Oystercatchers devote 45% of their feeding time to handling, this yield relation (depending on the observation that large mussels are taken nearly as fast as small ones) is of decisive importance. This is not the place to consider this relation in detail, but the striking difference between the 1973 and 1980 data on the size distribution of mussels on offer, with a corresponding shift in size of mussel taken by the Oystercater (Figure 1), calls for comment. The known relation between shell size and yield (Figure 2B) allows computation of the contri- bution of the various size classes of mussel towards the intake rate actually achieved, and with this reconstruction derived from the distribution of shell sizes taken by the Oystercatcher we can quantify the repercussions to the intake rate if mussels at the top or bottom of the distribution would have been ignored.

Intake rate is highly sensitive to mussels from the upper end of the size distribution, and were the Oystercatcher to ignore these, then the intake rate decreases markedly. At the other end of the scale, the Oystercatcher should accept each mussel as long as the yield (intake during handling) is equal to or greater than the average intake rate experienced (intake during total foraging time, i.e. during handling and searching). Viewed from this perspective, the rejection threshold for 1973 corresponds to a minimal size class of 38 mm (because the mean intake rate of 5.68 mg per second of feeding time, see Figure 2A, corresponds to the yield during handling of the 38 mm mussel, see Figure 2B). Acceptance of mussels below this threshold would lower the intake rate, and indeed this occurred hardly at all (Figure 1). The 1980 observations suggest a mean in- take rate of 1.72 mg/second, bringing the acceptance threshold down to ca. 23 mm, which is close to the observed lower limit (Figure 1). The contrast between the 1973 and 1980 seasons brings us to consider how the Oystercatcher decides when a food resource is no longer profitable to exploit, i.e. which criteria does the bird use to determine the lowest mean intake rate acceptable? Although we have

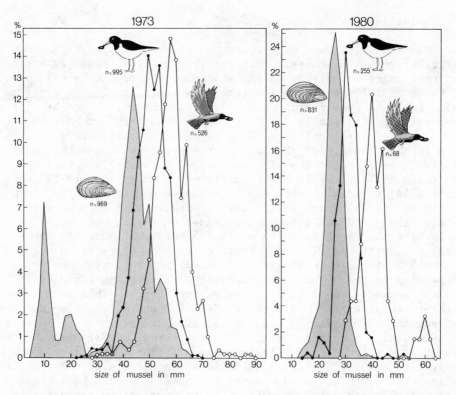

Figure 1. Size selection by Oystercatcher and Hooded Crow preying
on mussels (percent frequency distribution per 2 mm class, sample
size indicated). Shells opened by Oystercatchers (solid dots) are
compared to the mussel population on offer (stippled) based on
collections between August and November 1973 and in November 1980.
Hooded Crows transported the mussels to the sea-wall 1-2 km distant
where they cracked them (open dots, sampled November 1973 - January
1974 and November 1980).

side-stepped this question in the above procedure, it is relevant
to note that the main alternate prey for this area (Cerastoderma and
Macoma) are exploited down to a mean intake rate similar to that for
mussel feeders relying on sizes in the range 25-35 mm (Hulscher, in
preparation).

   The Hooded Crow, Corvus corone cornix, faces the added cost of
transporting the mussels to the sea-wall for extraction, and only
takes extremely large mussels (of a size we rarely encountered our-
selves in our sampling programme, see Figure 1), a nice example of
"central place foraging" (Orians, 1980).

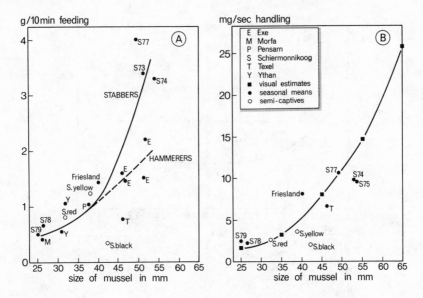

Figure 2(A) Feeding rate (g AFDW per 10 min feeding) in relation to average size (length in mm) of mussel taken. Oystercatchers on Texel (T) and Schiermonnikoog (S, year as shown) and the captive birds "red" and "yellow" were stabbers; "black" and the majority of Oystercatchers at other sites were "hammerers". Technique affects intake rate in the larger size classes (lines drawn by eye).
(**B**) Yield (mg AFDW) per second for handling for mussels of different size. Sources for both: Exe (Goss-Custard, pers. comm.); Frisian coast (Hulscher, unpublished); Pensarn and Morfa (Drinnan, 1957); Texel (Koene and Drent, in prep.); Ythan (Heppleston, 1971; size of mussel derived from flesh weight data). Visual estimates of shell size from Koene (Texel), line drawn by eye.

Feeding Rate in Relation to Bird Density

     Tidal variation in capture rate (Figure 3) may be explained by variation in the number of mussels gaping and hence susceptible to capture by Oystercatchers utilizing the stabbing technique. Therefore, one might expect the birds to concentrate on mussels covered by a film of water and thus to follow the tide-line, starting on the high-est parts of the mussel banks and working down to low-lying clumps as the tide recedes, but as shown in Figure 3B this did not occur on a large scale. Oystercatchers, individually recognizable, generally remained within a home range of less than 0.5 ha throughout the low water period. Detailed observations on behaviour of the prey in relation to the tidal cycle are called for before the trend in capture rate can be interpreted fully, but by selecting the period roundabout low water for further analysis, the greater part of the tidal varia-tion could be excluded. The next step was to examine capture rates

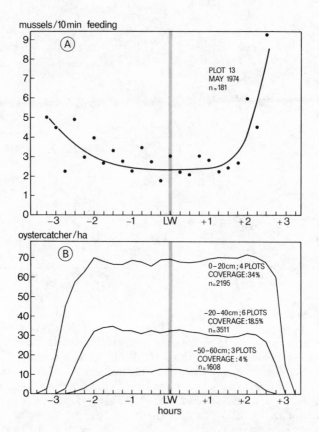

Figure 3(A)   Intake rate (number of mussels taken per 10 min feeding)
in the course of tidal cycle (plot 13, immersion time 6 hr). LW =
time of low water.
(B)   Average density (Oystercatchers per ha plot) given separately
for three height categories of the mussel bed (given in cm in relation
to mean sea level = NAP). The highest portions of the mussel bed had
a higher rate of coverage, hence the greater Oystercatcher density.
Note that after each sector becomes available numbers are constant,
implying that subsequent movements are minimal. Sample size (15 min
counts) indicated (May-November).

within this restricted period in relation to the number of birds
present (Figure 4) and indeed feeding rate decreased as the density
of the Oystercatchers went up, an effect found on Schiermonnikoog
as well as Texel. As Oystercatcher densities increase, the proportion of mussels stolen by other Oystercatchers and by Herring Gulls,
Larus argentatus (at Texel) or Common Gulls, Larus canus (at Schiermonnikoog) also rises, but in addition an interference effect of the
Oystercatchers themselves contributes to the decline in capture rate,
as could be verified by manipulating densities of semicaptive birds

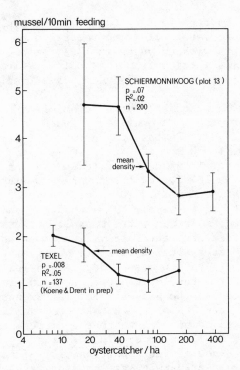

Figure 4. Density-related intake rate of Oystercatchers on a mussel bank near Schiermonnikoog (plot 13, August–November 1973; coverage 50%) and near Texel (September 1976 – April 1977; Mussel coverage 47%). Tidal effects do not obscure the density relation (see text). Statistical significance (p) and amount of variance explained (R^2) according to one-way analysis of variance is entered for each study area. For the density classes distinguished, standard errors are indicated. Although at a given density, intake at Texel is only one-third that observed at Schiermonnikoog, the "average bird" at Texel experienced an intake rate half that typical for Schiermonnikoog (see arrow connecting mean bird densities for each study area). Note that the abscissa represents hectare plots for which the mussel cover-age is c. 50%.

on the mussel bed and by the observation that the feeding activity of wild birds was depressed through the addition of model Oyster-catchers on the mussel beds (Koene and Drent, unpublished).

Admittedly, only a small part of the variance in capture rate was accounted for by the number of birds present, but fortunately for us, water level, which as already mentioned could account for a larger part of the variance, was independent of bird density in these data. Individual differences between Oystercatchers give a large

variation in the observed feeding rate: for example, parents feeding
their young (n=6) achieve collecting rates at, on average, 1.7 times
the level for adults without young (n=179) feeding at the same spot
and at the same time. Aside from this complication, bird density of
itself depresses capture rates of individuals feeding at the same time.

Although the shape of the curve relating capture rate to bird
density is similar, the much lower level found on Texel demands
explanation (Figure 4). In both cases the size of mussels captured
was about the same (46.2 mm for Texel compared to 50.5 mm for Schiermonnikoog), both mussel banks were situated at the same elevation in
relation to mean sea level, and the coverage by mussels was also
closely similar (47% and 50%). There was, however, a striking difference in the density of mussels (1790/m^2 on Texel and 3020/m^2 for plot
13, Schiermonnikoog). In fact, the mussels large enough to be attractive to the Oystercatcher were nearly five times as scarce on Texel
as on plot 13 (489 as against 2223/m^2 respectively for all mussels
larger than 40 mm). The low mussel density on Texel is the most
likely explanation for the low capture rate, and in addition, more
immature birds frequent the bed at Texel (C. Swennen, pers. comm.)
than on plot 13, where 70% of the birds were adult and this difference can be expected to contribute to the difference in capture rate
(Norton-Griffiths, 1967), although in our data capture rate in these
age categories does not differ significantly. In any case, search
time per mussel was nearly twice as long for Texel as on plot 13
(222 sec as contrasted to 116 sec), whereas handling times were nearly equal (77 sec versus 66 sec).

Dispersion Rules

Since we have seen that a high Oystercatcher density has an
inhibitory effect on the mussel capture rate, it is to be expected
that as a mussel bank becomes more crowded, birds might achieve a
higher feeding rate by opting to settle elsewhere, seeking compensation for less abundant food supplies and longer travelling times by
selecting less crowded mussel beds. As shown in Figure 5, a buffer-
ing effect on feeding areas relatively near to the roost was indeed
found. At periods of low bird numbers about 12% of the Oystercatchers
present fed on the preferred mussel banks A and B (comprising 12 ha
mussel together, i.e. only 0.3% of the area available), but this
proportion declined to 8% at the maximal numbers. The proportion of
Oystercatchers which reached the decision to fly on to mussel banks
5-7 km from the roost thus increased with increasing bird numbers in
the area. Apparently on account of the distance to the roost, bank
B was settled less densely (note that Oystercatchers utilization is
plotted as birds per hectare actually covered with mussels). The
difference in utilization of bank A between 1971 and 1973 might have
to do with the density of cockles Cerastoderma edule, an important

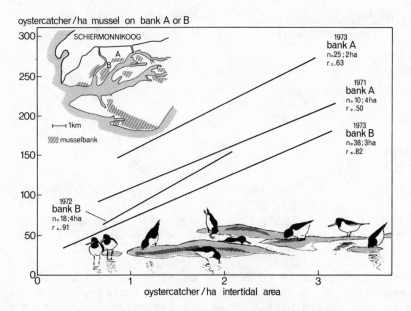

Figure 5. Feeding density of Oystercatchers per ha mussel bed surface (derived from counts of birds per 1-ha plot and data for mussel
coverage in the plot) on mussel bank A and B (see inset map) as
related to the average feeding density on the entire intertidal
area (c.4000 ha SW of Schiermonnikoog shown on map) as derived from
counts on the high water roosts on western Schiermonnikoog for the
same day (May-November).

alternative prey, which was relatively more abundant in 1971 (Essink,
1978 and pers. comm.).

The next question is, whether such adjustments of density can
also be discerned within a mussel bank, when small-scale differences
in configuration of the food stock are considered (as indicated by %
coverage). A prerequisite to such an analysis is the selection of
data restricted to periods when Oystercatchers feeding out on the
mudflats on other prey were rare or absent; for that reason counts
made in spring when some birds utilized Macoma balthica were omitted.
Figure 6 shows that as the mean density on the mussel banks increased
from 10 to 50 Oystercatchers/ha, the number utilizing the clumps
(only present in plots where mussel coverage was less than 5%) went
up only 2-3 times, whereas on the actual bed (coverage above 20%)
the density increased sevenfold. From this we concluded that the
penalty to the newcomer of settling on the clumps is far higher than
on the more extensive bed, i.e. the behaviour of the birds provides
a buffer effect. Furthermore, if the data is converted to a datum
perhaps more relevant to the feeding Oystercatcher, the density of
birds per ha mussel surface available (Figure 7), the clumps are

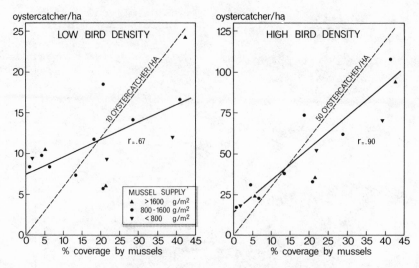

Figure 6. Importance of configuration of the mussel bed for foraging
Oystercatchers: bird density (per ha total surface, each point is
a mean for all counts within 30 min of low water) in relation to
mussel coverage distinguishing tides with low bird density (left
panel, n=29, average density less than 20 birds per ha) from tides
with high bird density (n=26, average density 40-60 birds per ha).
Lines of regression shown (considering coverage irrespective of
density, see key), and for comparison the distribution expected if
the feeding density per unit surface mussel bed is equalized between
plots (broken line). According to our observations Oystercatchers
at all densities concentrated on mussels.

characterized by an extremely high utilization rate. Although a
small over-estimate of the apparent utilization of the clumps cannot
be excluded since birds counting in the clump sectors may occasionally
have captured other prey, the reality of a higher predation pressure
on the clumps was verified from the sampling data considered later
(presented in Figure 12). Interestingly, two other bird species like-
wise feeding on the mussel bank show heightened density on the clumps
when plotted with respect to mussel area available (Figure 7B: the
Common Gull depended almost entirely on stealing mussels from Oystercatchers and hence reflect Oystercatcher density, whereas the Herring
Gull harvested small mussels (less than 20 mm) and large shorecrabs
(larger than 40 mm) and robbed Oystercatchers only occasionally).
For all three species mussel coverage was the most significant factor
determining dispersion over the study plots, accounting for 68-82%
of the variance. Differences in prey density failed to account for
a significant part ($R^2 \leq 1\%$) of the variance when multiple regressions
were performed (prey density was taken as g/m^2 of mussels larger than

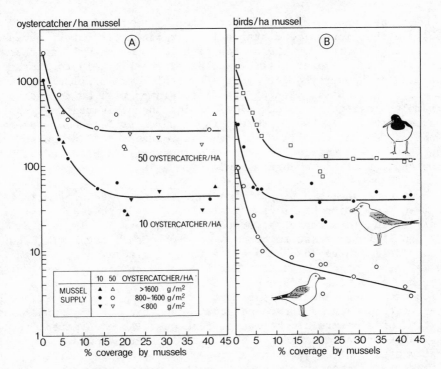

Figure 7(A)   Density of Oystercatchers (same data as Figure 6) replotted as birds per ha mussel bed surface, in relation to per cent mussel coverage in the plots (high and low density counts distinguished as in Figure 6).
(B)   Average density of Oystercatcher, Herring Gull, and Common Gull (from top to bottom) per ha mussel bed surface, in relation to per cent mussel coverage (n=109 for all species, low water counts for bank B, July-November 1973). All lines drawn by eye.

40 mm for the Oystercatcher, see key in Figure 7A, g/m^2 of mussels smaller than 20 mm for the Herring Gull, and simply density of Oystercatchers for the Common Gull).

A density-related change in dispersion can only be understood if the effect of bird density on the feeding rate is known, as has been shown for mussel beds earlier (Figure 4). Unfortunately, such detailed data are not yet available for isolated mussel clumps, and whether capture rate can serve as a cue in adjusting density depending on configuration of food supply must remain an open question, although we consider it likely. We do have data on the proportion of birds actively engaged in feeding, in relation to mussel bed configuration, and with the help of these data we can further develop the argument.

Considering the effect of bird density on feeding rate, it is plausible that interbird distance is a critical parameter, because interference effects on the search path followed as well as the probability that a prey will be stolen are both likely to increase depending on how many other birds are within some critical distance (Vines, 1980). Birds joining others at clumps will be forced to land extremely close to birds already foraging, hence increasing density in areas with low mussel coverage is bound to produce a more pronounced effect on feeding rate than in areas of moderate to high mussel coverage, where the birds feed scattered over extensive beds and have more opportunities to avoid contiguity.

Our data on inhibition of feeding (Figure 8) support this idea. The mean proportion of birds actually feeding decreased with rising densities in plots where the mussel coverage was low (Figure 8, left panel) in contrast to plots with high mussel coverage (Figure 8, right panel), where no consistent effect of bird density emerged. The magnitude of the depression of feeding when the data are split into the categories 1-10% and 11-26% was the same, hence these categories are pooled here. Aside from bird density, the importance of other factors such as air temperature, condition of the mussel, and duration of exposure of the mussel bank was also considered. Holding these three factors constant in a multiple regression procedure, the density effect became even more pronounced (see Figure 8, left panel), so there is no reason to fear that contaminating variables have led to a spurious correlation.

Equipped with these two sets of data on density effects (the inhibition of proportion of time spent feeding when clumps are considered (Figure 8) and the depression of capture rates on extensive mussel beds (Figure 4)) we can now provide an interpretation of dispersion rules for Oystercatchers feeding on mussels (Figure 6). As we see it, feeding rate (Figure 9A, here given in terms of intake per time present on the mussel bed, i.e. involving both proportion of time spent feeding, and capture rate during active search) is the variable the bird is attempting to maximize, and at the same time it may provide the proximal clue towards deciding whether to stay in a given site or not. In line with the heightened effects of crowding in the preferred sites (the clumps, see Figure 9C), the increment in density in this habitat as the total bird population goes up, is less than the expansion on the extensive mussel bed (Figure 9B, the distribution predicted if intake rate is maximized; by way of example the relative proportions of clump to mussel bed is here taken to be 50-50 but of course for every ration a prediction can be arrived at).

Predation Pressure

The decline in mussel stocks through the year is known from the sampling programme (Figure 10). Selective predation counteracts

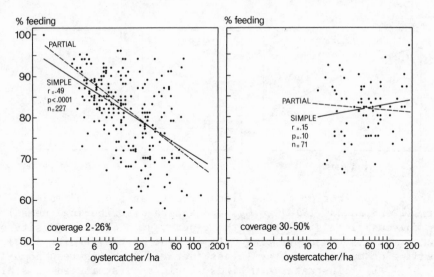

Figure 8. Proportion of Oystercatchers actually feeding in relation
to total density (birds per ha total surface) distinguishing the 9
plots with low mussel coverage (average ± SD 11 ± 8%) (left) from the
4 plots with high coverage (average ± SD 40 ± 9%) (right). Each dot
represents the mean for all 15 min counts for one low water period.
Lines of regression are shown uncorrected for other variables
(simple) and with other significant variables (immersion time, mussel
condition, air temperature) partialed out (partial). For the low
coverage plots a marked influence of bird density is apparent. Note
that if density of Oystercatchers is expressed in relation to area
actually covered by mussels, bird density in the low coverage plots
actually exceeds that on the high coverage plots.

growth to such an extent that the expected shift to the right in the
frequency diagrams as the season progresses fails to become apparent.
The number of mussels removed from the plots by the Oystercatchers
was estimated as follows. The number of Oystercatchers feeding per
low water period during the day is known from the 15 minute counts
made from the tower. Since the time the various portions of the
mussel bank were exposed varied from 4 to 6 hr (see Fig. 3B), predation pressure can be expressed by multiplying the number of Oystercatchers feeding by the feeding time, providing bird-hours per low
water period for each of the ha plots (Figure 11). Here again, the
buffer effect met with earlier manifests itself when fall, with its
higher mean Oystercatcher population, is compared with summer. Counts
made with the aid of infrared binoculars in September 1973 and May
1976 revealed that at night the same number of birds were present
on the plots as during the daylight low water periods, hence feeding
density by night is taken as equivalent to the daytime counts.

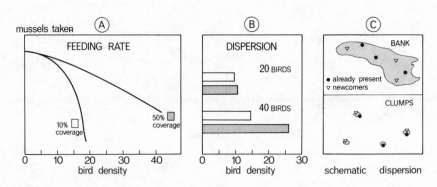

Figure 9. Application of the "ideal free distribution" model of
Fretwell and Lucas (1970) to Oystercatchers distributing themselves
over the mussel beds (A) shows the repercussion of bird density on
intake (mussels taken per time unit present on the feeding area)
depending on mussel coverage class (derived from Figure 4 combined
with 8); (B) illustrates how birds should be distributed to achieve
maximal intake rate for the average bird for two density levels
(total 20 and 40 birds) assuming equal total surface areas for the
two coverage classes; (C) visualizes the relatively heavier impact
of increasing bird density in scattered mussel clumps as distinct
from the mussel bed; hence the difference in rate of density increase
in these two habitat types (cf Figure 6).

        Intake rate is known for plot 13 by day. The intake rate for
nocturnal tides was measured in plot 13 in May 1976 again using
the infrared binoculars, and mussel intake was only 42% of the daylight value. This nocturnal depression is somewhat stronger than
captivity measurements indicate (31% depression, Heppleston, 1971;
37% depression, Hulscher, 1974) but in our field situation the Oystercatchers supplemented the mussel ration with some alternative prey
as well, hence the overall intake achieved by night was more closely
in line with the captive work than these figures indicate. Our intake figures are taken as 3.4 mussels per 10 min feeding by day and
1.4 mussels per 10 min feeding by night. The calculated number of
mussels removed by the Oystercatcher population is shown in Figure
12, along with the sampling data giving the total decrement. On
the isolated mussel clumps the decrease in the mussel population was
twice as heavy as on the mussel beds themselves, as a consequence
of the heavier predation pressure on what we have considered to be
the preferred sector. For plot 13 it is also possible to provide a
figure for the proportion of mussels present in summer 1973, that
were removed by the Oystercatchers in the course of the year until
summer 1974 (counts of Oystercatchers were resumed in April and May
of 1974, and from winter visits we concluded that the bird density
on the mussel area did not change). In all, the Oystercatchers took

some 40% of the mussels originally present (considering only the mussels larger than 40 mm).

On the mussel banks studied it took three years for the mussels to reach the size preferred by Oystercatcher (thus the spatfall of 1976 would become available to the Oystercatcher, i.e. cross the 40 mm threshold, either in 1978 or 1979). Following the successful spatfall of 1970 there was no measurable recruitment to the size class taken by Oystercatcher (see Figure 13). Even the heavy spat-falls of 1978 and 1979 have proven insufficient. A likely explana-tion for this failure of recruitment in ten successive years (1971 up to 1980 inclusive), is the severe predation on the first-year mussels by Herring Gulls, reducing the mussels remaining for Oyster-catchers more than two years later. Herring Gulls have increased in the study area recently. In August 1967 and 1968 they were lack-ing altogether on the mussel banks studied, whilst the Oystercatcher was present in the same density as in 1971-1973 (Roselaar, 1970). In August 1971 the Herring Gull occurred at a density of 7.1 per ha mussel, in August 1972 at 20.9 and August 1973 at 47.8. No detailed counts were made in the following years, but incidental counts indi-cate that the density of gulls has risen still further. The recent annual rate of increase of the Herring Gull population in the Netherlands as a whole amounts to 14% (Spaans, in Teixeira, 1979).

Figure 14 shows how Oystercatchers reacted to the failing recruitment and hence decreasing mussel densities. The feeding den-sity of Oystercatchers on the mussel bank declined in the course of the years, although the total number present on the roost remained the same (compare Figure 14B and 14D). In the course of the years more and more Oystercatchers were observed feeding out on the mud-flats, where they fed upon Cerastoderma edule, Macoma balthica, and Scrobicularia plana as well as the polychaete Nereis diversicolor. Oystercatchers remaining on the mussel bank however, achieved the same feeding rate in the first five years of the study (Figure 14C) despite the drastic decline in prey density (Figure 14A). After 1977, when few mussels above 30 mm were present, mussels of a small-er size were taken, and even though capture rate increased this was insufficient to compensate for the smaller flesh yield and intake rate in g AFDW declined (Figure 14C) to reach levels similar to those found by Heppleston (1971), and Drinnan (1957) for Oyster-catchers when preying on small mussels (Figure 2A).

Conclusions

The finding that the presence of conspecifics influences feeding rate, i.e., intake depends on density of Oystercatchers as well as of the prey, in combination with the observed stability of intake rate achieved on a specific site on the mussel bed studied during a long-term decline of mussel stocks over a period of many years, implies a continuous adjustment of feeding density to match the prevailing

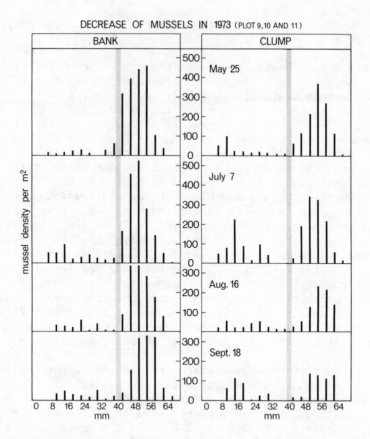

Figure 10.   Change in density of mussels per m² from May through September 1973 on mussel banks (left, extensive beds of at least 1000 m²) and mussel clumps (each less than 5 m²) per 4 mm classes (5-8 mm, 9-12 mm, etc.). The grey bar delineates the minimum size taken by Oystercatchers (see Figure 1).

conditions. This process of adjustment entails competition between individuals, and current work on marked individuals brings home to us how complex this is.

Not only is the distribution of birds over the available mussel banks involved, but clearly many individuals have switched to other prey as well. The wide spectrum of potential prey species for the Oystercatcher is well documented (Heppleston, 1971; Hulscher in Glutz von Blotzheim et al., 1975) and it will be recalled that the mussel bed in our study was most likely the most profitable food source in the local area at least until 1978. In our view, feeding rate experienced may serve as a proximal mechanism in bringing about the optimal degree of dispersion. We see no need to invoke less

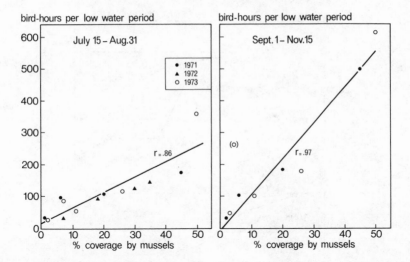

Figure 11. Bird-hours (time in hours the plot was uncovered multi-plied by the average number of Oystercatchers feeding per low water period) in relation to mussel coverage (July–August 4071 counts, September–November 2124 counts). The point for plot 12 in the fall (bracketed) is not included for calculations of the line of regression because most birds switched to feeding on moribund cockles in October.

direct assessment techniques such as the model of hunting by expecta-tion proposed by O'Connor and Brown (1976) who argued that Oyster-catchers utilizing cockles might use the accumulation of empty shells from previous exploitation, at "anvil" sites on the mudflat, as a cue to help guide them to the best sites remaining, the shell accumulation being used as a measure of resource depletion. Feeding rate, on the contrary, is sensitive both to prey availability and to the pressure of conspecifics, and provides a direct measure of profitability.

A remarkable feature of our data is the stability of absolute numbers of Oystercatchers in the study area through the years, and much more work, including observation at mussel banks further afield, and measurements of utilization of alternative prey, will be required before we can determine how total numbers are regulated by food supply. Interference among predators as here documented leads to distribute the birds more evenly among the available resources before local depletion reduces profitability, and we agree with Hassell (1978:95) that interference is a salient component of an efficient foraging strategy.

The finding that Oystercatchers deplete their prey locally is in line with other studies on this species (Goss-Custard, 1980).

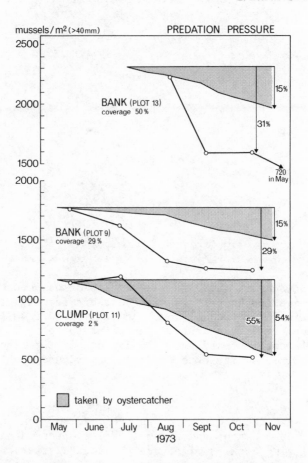

Figure 12. Disappearance of mussels larger than 4 cm in the course
of 1973 as deduced from sampling on the dates shown (open dots) and
portion explicable by predation of Oystercatchers (derived from data
collected in Figure 11 and 14C plus estimates of nocturnal predation,
see text). Oystercatcher impact on plot 11 (mussel coverage 2%) is
an overestimate, since some birds took other prey in the period
May-July. The predation pressure is on average 10% higher if the
mussels stolen by gulls are included.

The impact on the prey population exerted is by no means unusual when
compared with assessment of overwinter losses in other systems (see
Table 1).

     The demonstration of overfishing in our study is in contrast to
the earlier conclusion of Milne and Dunnett (1972) that the mussel
bed in the Ythan estuary was being cropped to a maximum so that no
net changes in size of the mussel population were occurring. Partic-

Table 1. Overwinter Impact of Birds on their Food Supply

Prey Organism	Predator	% prey population removed each winter	Authority
Intertidal invertebrates	Various waders	25–40%	eight studies summarized by Goss-Custard, 1980
Mytilus edulis	Oystercatcher	40%	this study
Dendroctonus rufipennis	Various woodpeckers	19–83%	Koplin, 1972
Dendroctonus frontalis (brood adults)	Various woodpeckers	18%	Kroll & Fleet 1979
Spiders, various spp.	Goldcrest	25%	Askenmo et al. 1977
Coleoptera laricella	Various woodpeckers	30%	Coppel & Cloan 1971 (in Otvos, 1979).
Podosesia syringae	Various woodpeckers	67–81%	Solomon in Otvos 1979
Ernarmonia conicolana	Tits	54–60%	Gibb 1960
Cydia pomonella	Various woodpeckers	34–63%	McLellan 1958
Cydia pomonella	Silvereye	53%	Wearing in Solomon & Glen 1979
Cydia pomonella	Tits	93–97%	Solomon et al. 1976

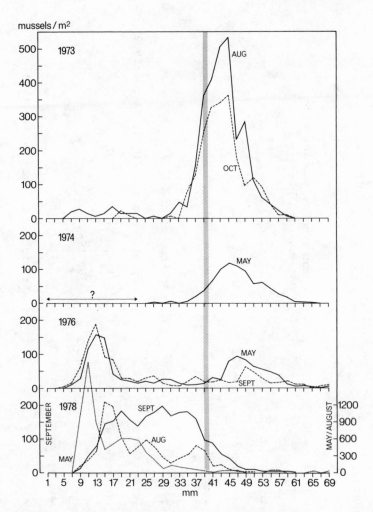

Figure 13. Density profiles of the mussel population of plot 13
through the years (class interval 3-4, 5-6, 7-8 mm etc.). Grey bar
delimits the minimal size of mussels taken by Oystercatchers (except
1978, when smaller mussels were also accepted). Although mussels
below 20 mm were not sampled in 1974, it is known that they were
rare. The decline in density during the 1978 season, a year of
heavy spatfall, is extremely sharp (note the two scales used).

ularly when studying intertidal organisms with their notoriously
erratic rates of spatfall, one should be alive to the possibility
that bird predators may indeed deplete the food resource to the
extent that it is no longer exploitable. In addition to the effects
of Herring Gulls and Oystercatchers in unison on mussel stocks as
presented here, there is now evidence for a similar local disappear-

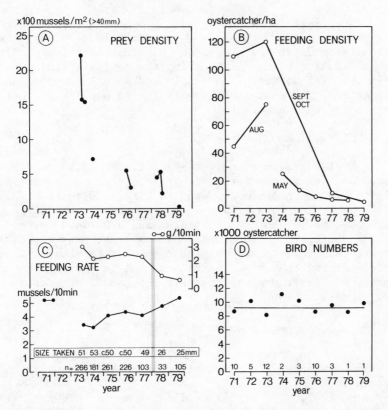

Figure 14. Overview of events on the mussel bed plot 13 through the years (A) number of mussels larger than 40 mm (see Figure 13), dots within one year are connected, (B) number of Oystercatchers in the plot; each point is the mean of several hundred counts, (C) intake rate (lower segment, mussels per 10 min feeding time) and in combination with data on average size taken the feeding rate (upper segment, g AFDW ingested per 10 min) can be derived with the aid of the length/weight relationships of the mussel populations concerned. Data for May (except 1973: August-September and 1979: September) n = number of 10 min observation periods, stippled bar calls attention to shift in prey size taken, (D) number of Oystercatchers present at the high water roosts for the SW Schiermonnikoog study area (August-November, number of counts indicated) corrected for seasonal trend within this period.

ance of the clam <u>Mya arenaria</u> due to predation by Oystercatchers on the second year clams and by Curlews on later stages (Zwarts, in progress).

We consider it likely that the case history of over-winter depletion of food stocks given here is representative for the

Waddenzee intertidal fauna as a whole. Competitive spacing as we
have observed it, delays but does not prevent local disappearance
of the food stock. We are convinced that the intertidal habitat
provides an ideal setting for tackling the controversial question
of the interaction between food supply and animal numbers, and hope
that current work on individually marked bird populations will
provide some of the answers.

ACKNOWLEDGEMENTS

It is a pleasure to acknowledge our debt to J.B. Hulscher, who
introduced us both to the world of Oystercatcher many years ago, and
participated in the programme presented here. We thank P. Koene for
so readily providing data from his studies on Schiermonnikoog and
elsewhere, B. Ens for helpful comments, and J.D. Goss-Custard for
access to his unpublished observations. A great many students took
part in the observations and we hope they will be pleased to see how
the pieces fit together. M. Zwarts assisted in writing the computer
programmes, D. Visser prepared the figures, and J. Zwarts provided
the bird vignets.

REFERENCES

Askenmo, C., A. von Brömssen, J. Ekman and C. Jansson, 1979. Impact
    of some wintering birds on spider abundance in spruce. Oikos
    28: 90-94.
Drinnan, R.E. 1957. The winter feeding of Oystercatcher (Haematopus
    ostralegus) on the edible mussel (Mytilus edulis) in the Conway
    Estuary, North Wales. Fishery Invest., London, Ser. II, 22:
    (No. 4).
Essink, K. 1978. The effects of pollution by organic waste on macro-
    fauna in the eastern Dutch Wadden Sea. Neth. Inst. Sea Res.
    Publ. Ser. 1-1978.
Fretwell, S.D. and H.L. Lucas, Jr. 1970. On territorial behavior
    and other factors influencing habitat distribution in birds.
    I. Theoretical development. Acta Biotheoretica XIX (1): 16-36.
Gibb, J.A. 1960. Populations of tits and goldcrests and their food
    supply in pine plantations. Ibis 102: 163-208.
Glutz von Blotzheim, U.N., K. Bauer and E. Bezzel, 1975. Handbuch
    der Vögel Mitteleuropas VI. Akademische Verlagsgesellschaft.
Goss-Custard, G.C. 1976. Variation in the dispersion of redshank,
    Tringa totanus on their winter feeding grounds. Ibis 118: 257-
    263.
Goss-Custard, J.D. 1977. The ecology of the Wash. III. Density-
    related behaviour and the possible effects of a loss of feeding
    grounds on wading birds (Charadrii). J. Appl. Ecol. 14: 721-739.
Goss-Custard, J.D. 1980. Competition for food and interference among
    waders. Ardea, in press.
Hassell, M.P. 1978. The dynamics of arthropod predator-prey systems.
    Princeton University Press.

Heppleston, P.B. 1971. The feeding ecology of Oystercatchers (Haematopus ostralegus L.) in winter in northern Scotland. J. Anim. Ecol. 40: 651-672.

Hulscher, J.B. 1974. An experimental study of the food intake of the Oystercatcher Haematopus ostralegus L. in captivity during the summer. Ardea 62: 155-171.

Hulscher, J.B. 1976. Localisation of cockles (Cardium edule L.) in darkness and daylight. Ardea 64: 202-310.

Kluijver, H.N. and L. Tinbergen, 1953. Territory and the regulation of density in Titmice. Arch. Néerl. Zool. 10: 265-286.

Koplin, J.R. 1972. Measuring predator impact of woodpeckers on spruce beetles. J. Wildl. Mgmt. 36: 308-320.

Kroll, J.C. and R.R. Fleet, 1979. Impact of woodpecker predation on over-wintering within-tree populations of the southern pine beetle (Dendroctomus frontalis). In: J.G. Dickson, R.N. Connor, R.R. Fleet, J.A. Jackson and J.C. Kroll (eds.) The Role of Insectivorous Birds in Forest Ecosystems. Academic Press.

McLellan, C.R. 1958. Role of woodpeckers in control of the Codling Moth in Nova Scotia. Can. Ent. 90: 18-22.

Milne, H. and G.M. Dunnett, 1972. Standing crop, productivity and trophic relations of the fauna in the Ythan estuary. p. 86-106 In: R.S.K. Barnes and J. Green (eds.) The Estuarine Environment. Applied Science Publishers, London.

Norton-Griffiths, M. 1967. Some ecological aspects of the feeding behaviour of the Oystercatcher, Haematopus ostralegus on the edible mussel, Mytilus edulis. Ibis 109: 412-424.

O'Connor, R.J. and R.A. Brown, 1976. Prey depletion and foraging strategy in the Oystercatcher Haematopus ostralegus. Oecologia 27: 75-92.

Orians, G.H. 1980. Some adaptations of marsh-nesting blackbirds. Princeton University Press.

Otvos, J.S. 1979. The effects of insectivorous bird activities on forest ecosystems: an evaluation. In: J.G. Dickson, R.N. Conner, R.R. Fleet, J.A. Jackson and J.C. Kroll (eds.). The role of insectivorous birds in forest ecosystems. Academic Press.

Roselaar, C.S. 1970. Een onderzoek naar de aktiviteit van wadvogels op Schier. Schierboek 4: 113-144.

Solomon, M.E. and D.M. Glen, 1979. Prey density and rates of predation by tits (Parus spp.) on larvae of Codling Moth (Cydia pomonella) under bark. J. Appl. Ecol. 16: 49-59.

Solomon, M.E., D.E. Glen, D.A. Kendall and N.F. Milsom, 1976. Predation of over-wintering larvae of Codling Moth (Cydia pomonella) by birds. J. Appl. Ecol. 13: 341-352.

Teixeira, R.M. 1979. Atlas van de Nederlandse broedvogels. Natuurmonumenten.

Vines, G. 1980. Spatial consequences of aggressive behaviour in flocks of Oystercatchers, Haematopus ostralegus L. Anim. Behav. 28: in press.

Zwarts, L. 1976. Density-related processes in feeding dispersion

    and feeding activity of Teal (Anas crecca). Ardea 64: 192-209.
Zwarts, L. in press. Intra- and inter-specific competition for space
    in estuarine bird species in a one-prey situation. Int. Orn.
    Congress, Berlin.
Zwarts, L. 1980. Habitat selection and competition in wading birds.
    In: C.J. Smit and W.J. Wolff (eds.). Birds of the Wadden Sea.
    Balkema, Rotterdam.

FACTORS AFFECTING THE OCCUPATION OF MUSSEL (<u>MYTILUS</u> <u>EDULIS</u>) BEDS

BY OYSTERCATCHERS (<u>HAEMATOPUS</u> <u>OSTRALEGUS</u>) ON THE EXE ESTUARY, DEVON

J.D. Goss-Custard, S.E.A. Le V dit Durell, S. McGrorty,
C.J. Reading, R.T. Clarke

Institute of Terrestrial Ecology
Furzebrook Research Station, Nr. Wareham
Dorset, BH20 5AS

INTRODUCTION

Between two and three thousand oystercatchers, <u>Haematopus</u> <u>ostralegus</u>, feed on the Exe estuary, South Devon, between August and March and most are adults. A few hundred immatures are present in summer when the adults are breeding in northern Britain, Norway and Holland. Most adults on the Exe specialise in feeding on mussels whereas many immatures concentrate on other prey (Goss-Custard et al., 1980). Accordingly, few oystercatchers eat mussels between March and August but, thereafter, there is a rapid increase in the numbers doing so as the adults return.

The first adults to return encounter almost empty feeding areas so there is unlikely to be much competition for the best places. Most birds would, therefore, be expected to feed in the optimum sites. Correlations at this time between bird density and various features of the mussel beds may reveal those characteristics of the feeding areas which oystercatchers prefer. As the population increases, bird density on the preferred areas may eventually reach a limit so that an increasing number feed in, what are by definition, the less preferred parts of the shore (Zwarts 1974; Goss-Custard, 1977a,b). Consequently, bird density may become correlated with other features of mussel beds as the less preferred places are occupied.

This paper describes the sequence in which newly arrived oyster-catchers occupy the mussel beds and look for correlations between bird density and various features of the mussel beds at different population sizes. Our present understanding of their feeding strategy is then summarised.

MUSSELS AND OYSTERCATCHERS ON THE EXE

Details of methods are omitted to save space, but most are
described in Goss-Custard et al., (1980). The remaining techniques
are described in full in several papers now in preparation.

Thirty-one mussel beds were identified on the basis of the
density and size of mussels there. Some were too small to warrant
detailed study, but most were sampled each September. As will
become clear, the beds also varied in several other respects which
might affect oystercatchers and the appropriate measurements were
made at various stages in the study. Taken together, the mussel
beds provide a highly variable food supply for the oystercatchers to
use.

As over 85% of the oystercatchers on mussel beds eat mussels
(Goss-Custard et al., 1980), the numbers on the beds approximate
to the population actually eating this prey. Each bed was counted
regularly during late summer and autumn. The population on mussels
built up in a similar way in each year of the study (1976-1978) with
the numbers increasing almost six-fold from about 330 to 2000 between
July and October. When these studies were done, most birds roosted
at high water at Dawlish Warren at the mouth of the estuary.

SEQUENCE OF OCCUPATION OF THE BEDS

The distributions of the birds over all beds on two representative counts are summarized in Fig. 1. On each count, the beds were
ranked according to the proportion of the mussel-feeding population
found there. The beds were then arranged in ascending order of
rank and the proportion of the population on them accumulated. If
all the birds occurred on one bed, the curve would form a right-
angle. If the birds were distributed equally amongst all beds, it
would be a straight line with a slope of 100/31.

Two-thirds of the birds occurred on just two adjacent beds on
the low count (A) so the curve approximates to a right-angle. The
curve is much flatter in the high count (B) as the birds spread out
over the estuary. But partly because the beds vary considerably in
size, two-thirds of the birds still occur on only eight of the
largest beds. Therefore, the results of this count are also expressed
as cumulative densities rather than numbers (C). Though less marked,
the relationship is still curved because the density of oystercatchers
varies considerably between beds. While the birds spread out from
two highly preferred beds as the population size increases, their
distribution is still not uniform.

The tendency to spread out can also be illustrated with data
from a small sample of beds. Fig. 2 shows (A) the proportion of
the total mussel-feeding population, and (B) the numbers, occurring

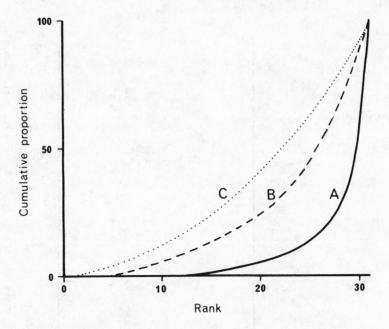

Figure 1. The distribution of oystercatchers over the 31 mussel beds of the Exe on a low count (A), when 331 birds were present, and on a high count (B and C) when 1989 birds were present. The proportion of the total number of birds is accumulated in order of bed rank in A and B, whereas densities are accumulated in C.

on two pairs of beds. The first pair, the preferred beds, have about half of the birds initially but the proportion decreases rapidly as the population increases. In contrast, the proportion on the second pair starts very low but increases as the population rises and are by definition, less preferred beds. Most of the beds follow the pattern shown by the second pair, although some are intermediate between the two extremes.

Although the beds are occupied in a fairly clear sequence, it is not a strict step-wise succession with the preferred beds filling up first before birds go elsewhere. Beds are occupied simultaneously with numbers on both pairs continuing to increase as the birds spread out over the estuary (Fig. 2B). In no case it a clear maximum density attained, in contrast to other studies on waders (Goss-Custard,1977a,b). If birds do indeed spread out because of competition for food or space on the preferred areas, it is because of a gradual increase in resistance rather than because the beds have filled up with birds and the excess then have to go elsewhere.

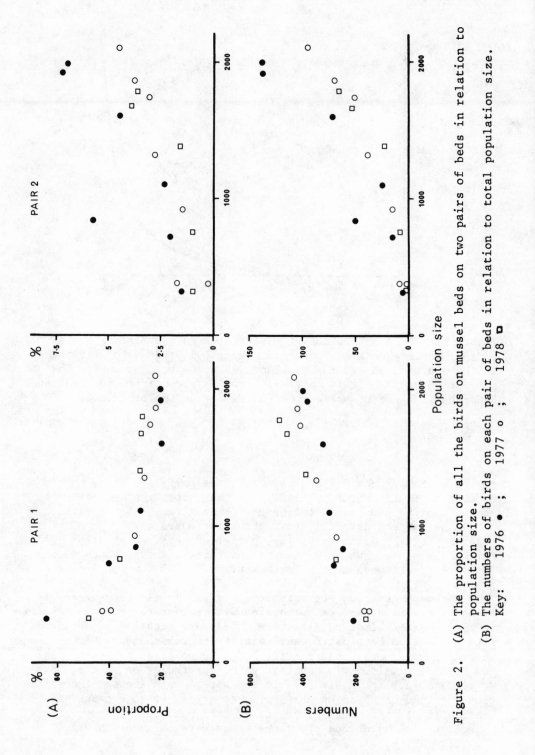

Figure 2.   (A) The proportion of all the birds on mussel beds on two pairs of beds in relation to
            population size.
            (B) The numbers of birds on each pair of beds in relation to total population size.
            Key: 1976 ● ;   1977 ○ ;   1978 □

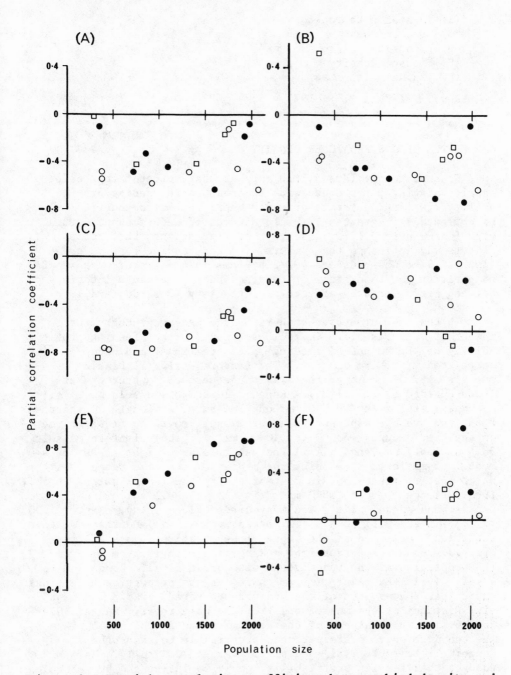

Figure 3. Partial correlation coefficient between bird density and each of 6 mussel bed factors plotted against population size.
(A) Shell thickness, ventral

continued ...

(B)   Distance to roost
(C)   Bed softness
(D)   Bed size
(E)   Mussel density
(F)   Mussel size

Key: 1976 ● ;   1977 ○ ;   1978 ▢

CORRELATES OF OYSTERCATCHER DENSITY

     Present theory (Krebs,1978) argues that, other things being
equal, foraging birds congregate in areas where feeding profitability (i.e. the net rate at which energy or scarce nutrient can
be collected) is greatest. Field studies of waders support this
view (Goss-Custard,1970, 1977a,b) but also suggest that social
factors may influence the numbers of birds feeding in some areas
(Zwarts 1974; Goss-Custard 1977a,b;Vines 1976), either because of
competition for food items or feeding space or because interference
between birds directly affects profitability (Goss-Custard,1976).

     In the case of oystercatchers on the Exe, we thought that a
number of features of the mussel beds could affect feeding profitability or the amount of feeding space available, and hence be
correlated with bird density. The factors were: (i) Density of
mussels over 30 mm long, the size range eaten by oystercatchers
(Goss-Custard et al., 1980). Range: 10 to 680 mussels m^{-2}. (ii)
Average mussel size over 30 mm long, measured as length to the power
2.85, the average exponent relating mussel length to flesh volume
on the Exe. Range: 33000 to 110000 mm$^{2.85}$, equivalent to mussels
38.5 and 58.7 mm long. (iii) The ash free dry weight (AFDW) of a
mussel 55 mm long, this being the median of the size range from
which oystercatchers obtain most of their food (Goss-Custard et al.,
1980). Range: 350 to 1840 mg. (iv and v) Shell thickness on the
ventral and dorsal side where most birds gained entry. Range: 1.00
to 1.38 mm and 0.81 to 0.97 mm respectively. (vi) Softness of the
substratum, measured as the depth of penetration of a steel rod
dropped from a standard height. Range: 3.8 to 18.2 cm. (vii)
The proportion of the bed covered with brown algae. Range: 0 to
41.4%. (viii) Exposure time, measured as the proportion of a tidal
cycle that 50% or more of the bed was uncovered. Range 15 to 46%.
(ix) Distance of the mussel bed from the main roost. Range: 0.7
to 5.4 km. (x) Flatness of the bed, measured as the proportion of
the bed over which oystercatchers are visible to each other.
Range: 25 to 100%. (xi) Bed size. Range: 0.2 to 14.4 ha. We
thought that feeding profitability could be influenced by variables
(i) to (viii) by affecting either the rate of intake or expenditure
of energy. Exposure time was included in case birds preferred to
feed where most time was available. Bed flatness was thought likely
to influence how much birds could see each other and, therefore,

their rates of interaction. Bed size could affect the ability of birds to avoid disturbance from people yet still remain on the bed or, perhaps, the absolute number of birds present and so the degree of social facilitation attracting still more birds to the bed.

The data from some small beds were combined whereas other beds were ignored because no data on mussels were available. Sample size on each count varied between 21 and 23, which is too small for a multivariate analysis with 11 variables. A preliminary analysis was, therefore, carried out by combining data from all three years into five sets according to the total numbers of birds on mussel beds, viz 250-500, 500-1000, 100-1500, 1500-1750 and 1750-2100. Sample size then varied between 84 and 151. The results of a step-wise multiple regression on the data from each group, to be published in detail elsewhere, resulted in two variables not being selected at all (variables iii and vii) and three being chosen sporadically but with a sign opposite to that predicted (variables v, viii and x). Since these variables also contributed little to the overall r^2, only the remaining six were used in the analysis of the individual counts.

Possible changes in the effect of each of these factors with increasing population size were investigated by analysis of each count. The density of oystercatchers on the 21-23 beds was regressed against each variable in turn with the effect of the remaining five being held constant by partial correlation. The resulting partial coefficients were then plotted against population size (Figs. 3A-F), and the correlation between these coefficients and population size calculated (Table 1).

Table 1. (A) The correlation between partial correlation coefficients and population size, and (B) the mean values of those where the correlations were not significant.

	(A) correlations		(B) means	
	r	p	$\bar{x}$	P value of difference from zero
A. Ventral shell thickness	+0.108	NS	−0.341	<0.001
B. Roost distance	−0.327	NS	−0.357	<0.001
C. Bed softness	+0.656	<0.01	−	
D. Bed size	−0.509	<0.05	−	
E. Mussel density	+0.805	<0.001	−	
F. Mussel size	+0.651	<0.01	−	

The results suggest that some factors exert an effect at all
population levels whereas the effect of others may change. (i) The
coefficients for shell thickness (A) and distance to roost (B) varied
independently of population size. Mean values were, therefore, calculated and were significantly different from zero (Table 1). This
result is broadly confirmed by the analysis of the grouped data
referred to earlier, except that their effect did not reach statistical significance in the <500 count band. (ii) The coefficients
for bed softness (C) and size (D) decreased significantly as population size increased (Table 1) bud did not reach zero except in the
case of bed size at the highest population levels. The analyses of
the grouped data confirmed that bird density was negatively correlated
with bed softness at all population levels and positively with bed
size, except in the largest count band. (iii) The effect of mussel
density (E) increased rapidly as the population increased, a result
confirmed by the grouped data. There was no evidence that mussel
density affected bird density on low counts, but it had an increasingly marked positive effect as bird numbers rose. (iv) The coefficients for mussel size (F) increased significantly with population
size and changed from negative to positive, a result again confirmed
by the analyses of the grouped data. Thus, bird density responded
negatively to mussel size at the start but became positive as the
population increased.

These results are summarised in Fig. 4. When the population
is small, the birds are densest on the large beds which are firm,
situated near to the roost and where the mussels are small and thin-
shelled. But as the population builds up and the birds spread out,
density becomes correlated with somewhat different factors. The
main changes are that more birds now feed where mussels are large
and dense. Shell thickness, distance to roost, bed size and,
particularly, substrate softness retain their importance, although
declining somewhat in the case of the latter two factors.

CORRELATES OF FEEDING PROFITABILITY

With the exception of bed size, the variables that were correlated with bird density were originally included in the analyses
because they were thought likely to affect either, the rate of intake, or the expenditure of energy while feeding. What evidence is
there that they affect either?

The gross rate at which mussel flesh was consumed (ingestion
rate) was measured by standard observational procedures (e.g. Goss-
Custard, 1977a). Data were obtained on 14 beds from August to
October (N = 23). We could not see how bed size would affect ingestion rate, so only the remaining five variables which affected bird
density were tested, although AFDW was also included. The correlation
of each variable with ingestion rate was calculated with the effect
of the other factors held constant by partial correlation analysis.

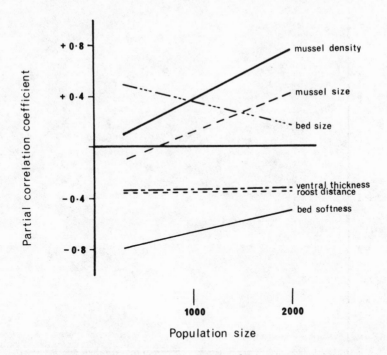

Figure 4. A summary of the relationships shown in Fig. 3. Regression lines are shown where significant correlations were obtained. Mean values are given in the remaining cases.

Although attempts are being made to measure how much energy waders expend in pecking, pacing and handling prey (Ferns et al., 1979), no studies have yet been published on oystercatchers eating mussels. The numbers of pecks and paces and the time taken to break into a mussel and swallow the flesh can be used as a crude guide to the energy cost of obtaining food. Though of limited use, they were measured on a total of 57 occasions during August to March and their correlation with features of the mussel beds examined.

Ventral shell thickness had the most consistent effect on profitability. Ingestion rate decreased as the mussel shells became thicker (partial correlation coefficient: -0.556, P <.02, Fig. 5) and the birds took longer to open each mussel (P <0.001). On the other hand, they made fewer pecks (P <0.01), presumably because mussels thin enough to break into were more scarce. But since pecking is likely to be inexpensive and pacing rate was unrelated to shell thickness, we can safely conclude that the net rate of energy intake was greater where shells were thin. The high densities of birds where shells were thin can, therefore, be interpreted as a

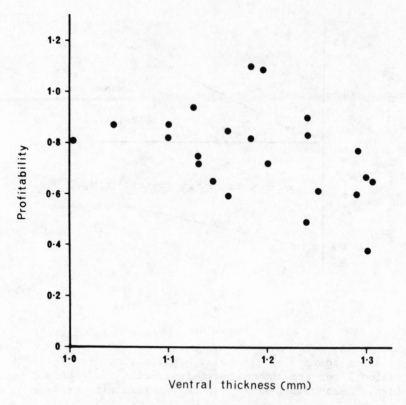

Figure 5.   Ingestion rate of oystercatchers against ventral shell thickness of mussel shells.

preference for beds with relatively high profitability.

Ingestion rate was not significantly correlated with any of the other variables tested.   There were poor partial negative correlations with shell size (-0.327) and substrate softness (-0.375) but the significance levels were low (P >0.10 in both cases).   Handling time increased with the sizes of shells present (P <0.001), but none of the other measures of expenditure correlated with either factor.

Mussel density, AFDW, distance to roost and bed softness were not correlated with ingestion rate.   Only AFDW, with which bird density is not correlated, appeared to have an influence on energy expenditure, because handling time was longer where mussels were heavy (P <0.001).   However, two of the factors may affect energy expenditure in ways which the crude measures employed could not detect.   It is highly likely that much more energy is used in walking on soft beds than hard ones because birds sink deeply into soft

mud. Breaking into shells may also be less expensive on hard beds
although, if so, we would expect handling time to be affected by
softness, and there is no evidence that it was. The higher
densities of oystercatchers on hard beds at all population levels
can, therefore, be interpreted as a preference for places where
energy expenditure through walking is low. This can also be said
of the preference for beds nearest to the roost because less energy
would be expended in reaching them, especially in strong winds.

To sum up, in late summer and autumn, profitability is greater
where shells are thin. Energy expenditure is probably least on
hard beds and on those near to the roost, and perhaps where mussels
are small. However, mussel density does not affect either the
rates of intake or expenditure of energy.

DISCUSSION

Pattern of Occupation of the Exe

The studies of Zwarts (this volume) and Vines (1976) strongly
suggest that social factors are involved as oystercatchers spread
out over mussel beds, and our unpublished findings on the Exe
support this view. We, therefore, assume here that social factors
are involved in the change in distribution as the population builds
up.

When the population size is low and competition presumably
minimal, oystercatchers prefer beds (i) where the shells are thin,
(ii) which are near to the roost, have a hard substrate and where
mussels are relatively small and (iii) which are large. A preference
for places of high profitability accounts for most of this because
gross intake rate is higher on beds with characteristics listed
under (i) and expenditure probably least on beds with characteristics
listed under both (i) and (ii). The preference for beds with small
mussels when population size is very low may be because the young
birds present at that time have to feed on easily opened mussels,
whereas the returning adults are probably better able to deal with
larger mussels. Oystercatchers prefer larger beds for reasons not
yet tested but unrelated to feeding profitability. Apart from this,
the birds' preferences in summer can be explained adequately in
terms of profitability.

As the population increases and spreads out, bird density
becomes increasingly positively correlated with mussel density and
size. Since neither factor is positively correlated with profitability, we conclude that these variables simply measure the amount
of feeding space available. More birds go to the less preferred
beds because that is where uncontested feeding space is to be found,
and more space is available where the mussels are dense and of the
large size taken by oystercatchers. In other words, the need to

find feeding space, whatever its quality, plays a bigger role in the
distribution of the oystercatchers as numbers rise.

Speculations on the Social Factors

Studies of marked individuals on the Exe suggest that most birds
remain on one mussel bed for long periods, and that they do most of
their feeding within a very small part of it (Goss-Custard et al.,
1980). Aggressive encounters are common and many seem to be concerned with competition for feeding space rather than single mussels.
If these attacks have the effect of reducing the local density of
oystercatchers, what costs and benefits does this have for the
attacking bird?

The main feeding cost is, presumably, that there will be fewer
birds to steal from. But there may be several benefits. Feeding in
one place throughout the winter may enable a bird to learn how best
to exploit it and attacking others may further improve efficiency.
During the winter, oystercatchers eat 20-30% of the mussels present
at the start (Goss-Custard et al., 1980) and, judging by the numbers
which they find but fail to break into, a high proportion of those
remaining may be unusable anyway. The birds may have a considerable
impact on their food and excluding some birds may significantly
reduce its decline. Alternatively, they may reduce the probability
that they themselves will be attacked by driving others away or by
establishing dominance over them. Or perhaps the mere presence of
another bird may reduce profitability through various kinds of
subtle interference for which there is field evidence for oystercatchers eating mussels from Holland (R. Drent pers comm) and the
Exe (unpublished information). In the latter case, birds may leave
a mussel bed because of their reduced profitability following
increased bird density rather than as a direct result of being
attacked too often, and the rates of aggression may be a consequence
not a cause of density which may be achieved by other means, such
as avoidance (Vines,1976). But whether the mechanism is for birds
to be driven away or just to leave an area when the profitability
becomes too low, studies of the effect of bird density on intake
rate may provide an explanation of the spreading out of the birds
as numbers build up and their increasing tendency to use feeding
areas of intrinsically poorer quality.

ACKNOWLEDGEMENTS

We are grateful to Drs. D. Jenkins, C. Milner and M.G. Morris
for comments on the manuscript.

REFERENCES

Ferns, P.N., MacAlpine-Leny, I.H. and Goss-Custard, J.D. 1979.
Telemetry of heart rate as a possible method of estimating

energy expenditure in the redshank, Tringa totanus (L.).
In: C.J. Amlaner Jr. and D.W. MacDonald (ed). A Handbook on
Biotelemetry and Radio Tracking. Oxford, Pergamon Press.

Goss-Custard, J.D. 1977a. The ecology of the Wash III. Density-
related behaviour and the possible effects of a loss of feeding
grounds on wading birds (Charadrii). J. appl. Ecol. 14: 721-
739.

Goss-Custard, J.D. 1977b. Predator responses and prey mortality in
redshank, Tringa totanus, and a preferred prey, Corophium
volutator. J. Anim. Ecol. 46: 21-35.

Goss-Custard, J.D., McGrorty, S., Reading, C.J. and Le V. dit
Durell, S.E.A. 1980. Oystercatchers and mussels on the Exe
estuary. Essays on the Exe estuary. Devon Ass. Special
Vol. 2.

Krebs, J.R. 1978. Optimal foraging: decision rules for predators.
In: J.R. Krebs and N.B. Davies (ed). Behavioural Ecology.
Oxford, Blackwells.

Vines, G. 1976. Spacing behaviour of oystercatchers, Haematopus
ostralegus, in coastal and inland habitats. Unpubl. Ph.D.
Thesis, University of Aberdeen.

Zwarts, L. 1974. Vogels van het brakke getijgebied, Amsterdam.

SURVIVAL OF SHOREBIRDS (CHARADRII) DURING SEVERE WEATHER: THE

ROLE OF NUTRITIONAL RESERVES

N.C. Davidson

Department of Zoology
University of Durham
South Road, Durham
DH1 3LE

INTRODUCTION

During winter, shorebirds face increasing difficulty in achieving sufficient rates of prey intake to satisfy their energy requirements, as a result of the deterioration in environmental conditions.
The effects of severe winter weather are twofold. Firstly, invertebrate prey species living in muddy or sandy substrata often become
less available to shorebirds, either because they become less active
and so less detectable, or by lying deeper in the substratum, mainly
in response to low temperatures, and also to high winds (Smith, 1975).
These changes in prey availability, and the responses to them by
shorebirds, are discussed by Evans (1979) and Pienkowski (1980a, and
this volume). The responses by shorebirds involve changes mainly in
foraging behaviour and/or in prey species taken (Goss-Custard, 1969;
Smith, 1975; Evans, 1976; Davidson, unpublished). The second, direct,
effect of severe weather on shorebirds is to increase the energetic
costs of thermoregulation during periods of low temperatures and high
winds. Gales can also affect shorebirds directly by preventing them
from feeding on their preferred feeding grounds (Evans, 1976; Dugan
et al., in press).

Other than changes in food and feeding behaviour, the main adaptation by shorebirds for overwinter survival is the storage of nutritional reserves. The main energy reserve in shorebirds (Evans and
Smith, 1975; Davidson, 1979; Dugan et al., in press), as in other birds
(e.g. Ward, 1969; Newton, 1969; Ankney, 1977), is fat. Muscle protein,
particularly the pectoral muscles, is also used as a nutritional
reserve (Kendall et al., 1973; Evans & Smith, 1975; Ankney, 1977).
Lipid reserves are metabolised whenever the rate of food intake is
insufficient to balance the rate of energy expenditure, whilst muscle

231

protein is metabolised only when little or no feeding is possible,
and is used primarily to provide amino-acids for protein synthesis
rather than energy (Evans and Smith,1975), except in the last stages
of starvation.

This paper discussed the role of lipid and protein reserves in
the survival of shorebirds through severe winter weather. It examines
three aspects of the role of nutritional condition: 1). the extent
to which different species of shorebirds use their nutritional
reserves during the same periods of severe weather, 2). the nutritional condition at which shorebirds die from starvation, and 3).
recovery after severe loss of condition, in relation to starvation
levels. Additionally, some further information on the weather conditions that affect the ability of different species to maintain energy
balance is discussed.

METHODS

Shorebirds have been collected at Teesmouth ($54^{\circ}37'N1^{\circ}12'W$) and
Lindisfarne ($55^{\circ}40'N1^{\circ}50'W$) in north-east England since 1971, during
studies of nutritional condition (Evans and Smith, 1975; Davidson,
unpublished), food and feeding habits (Smith, 1975; Evans et al.,1979)
and heavy metal pollution (Evans and Ward,unpublished). Only samples
collected during severe weather are discussed in this paper. Methods
of analysis for lipid and protein reserves, involving drying of
carcasses in vacuum ovens, followed by the extraction of lipids using
petroleum ether in Soxhlet apparatus, are described in detail by
Evans and Smith (1975) and Davidson (unpublished). Lipid indices
(i.e. lipid weight expressed as a percentage of total body weight)
of live samples of birds netted at Teesmouth have been estimated
from formulae relating wing-length and bill-length to lean weight
(Davidson unpublished).

Most published weights of shorebirds that had starved to death
have been given as total body weight. For comparison with 'normal'
winter lean weights, these have been converted by the subtraction
of 1% lipids (Marcstrom and Mascher, 1979; Baillie and Davidson,
unpublished). These lipids are unavailable for energy production
and lie, for example, in membranes.

Meteorological data have been obtained from several coastal
weather stations close to Teesmouth and Lindisfarne. Additionally,
a windchill factor (W) has been calculated from the formula $W = TV^{\frac{1}{2}}$,
where T is the mean daily temperature deficit below 20ºC, and V is
the mean daily windspeed in knots. This differs from the windchill
factor calculated by Dugan et al. (in press), who used a temperature
deficit below 10ºC.

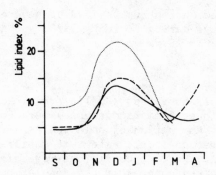

Figure 1. Normal winter patterns of lipid indices in adult shorebirds in north-east Britain. The three species shown are Bar-tailed
Godwit (—————) from Evans and Smith (1975), Dunlin (———————)
from Davidson (in prep.) and Grey Plover (·······) calculated from
total body weights in Dugan et al. (in press).

NORMAL WINTER LEVELS OF NUTRITIONAL CONDITION

    Lipid indices of shorebirds in north-east England (Figure 1)
rise during autumn and early winter to a mid-winter peak. This
peak may be in December (Bar-tailed Godwit Limosa lapponica) or in
December and January (Dunlin Calidris alpina, Knot C.canutus, Grey
Plover Pluvialis squatarola). Peak lipid indices are 12-15% in
most sandpipers (Scolopacidae). In plovers (Charadriidae) peak
lipid indices are higher, varying from 22% in Grey Plover (Figure 1)
to at least 25% in Golden Plover Pluvialis apricaria. Lipid indices
decline after mid-winter to about 7% in all species by early March.
In spring, lipids are rapidly accumulated before migration to the
breeding grounds, the timing of this spring increase differing
between species (Figure 1). Juveniles follow similar seasonal changes
in fat levels to adults of the same species.

    In adults in north-east Britain, pectoral muscle size (used as
a protein reserve index) is stable throughout the winter in Bar-
tailed Godwit (Evans and Smith, 1975), Dunlin and Knot, before rising
in spring in advance of migration and breeding (Davidson, unpublished).
The pectoral muscles of juvenile shorebirds are smaller than those
of adults in early winter, but have grown to adult size by January.
The pectoral muscles of plovers form a greater proportion of the
total lean weight (7-8%) than those of sandpipers (5-6%). The
absolute size of pectoral muscles in relation to the area of muscle
attachment is also larger in plovers than sandpipers (Davidson,
unpublished).

LOSS OF NUTRITIONAL CONDITION DURING SEVERE WINTER

    Information on the nutritional condition of several shorebird

species was collected during two periods of severe winter weather:
a period of seven days during February 1978, and the prolonged
severe weather of early 1979.

February 1978

Samples of three shorebirds (Dunlin from Teesmouth, Bar-tailed
Godwit and Golden Plover from Lindisfarne) were collected at the
beginning and end of the period of cold weather (Table 1), and
showed marked differences in the extent of mobilisation of their
nutritional reserves. Weather conditions were similar in both areas,
and there were no marked changes in the population sizes of Dunlin
or Bar-tailed Godwit during this period. The 'before -' and 'after
severe weather' samples of these two species are, therefore, thought
to be from the same populations. Golden Plover population changes
are discussed below.

Adult Dunlin used neither lipid nor protein during six days
of severe weather. In juvenile Dunlin, although the decrease in
mean lipid index is not significant (because of high variance in
both samples), birds used part of their lipid reserves since several
had lower lipid indices on 15 February (ranges were 5.3 - 11.2% on
9 February, 3.5 - 11.9% on 15 February).

In Bar-tailed Godwit, the lower lipid and protein levels on
15 February are consistent with the use of some nutrient reserves,
but sample sizes are very small and none of the differences are
significant.

In contrast, Golden Plover drew extensively on their nutrient
reserves during the same period, both lipid and muscle protein levels
decreasing significantly (Table 1). This extensive use of nutrient
reserves was a result of heavy snowfall at the start of the period
making pastures, the preferred habitat of Golden Plover (Fuller and
Youngman,1979), untenable. Numbers of Golden Plover on Lindisfarne
decreased early in the severe weather (pers. obs., J. Brigham pers.
comm.). Birds possibly moved south or south-west, as recorded
during previous severe weather by Ash (1964) and Dobinson and Richards
(1964). Although the sample taken at the end of the severe weather
may, therefore, be representative of only part of the population
present at the start of the cold spell, the Golden Plover collected
on 14 February had undoubtedly mobilised nutrient reserves during
the severe weather since the ranges of values were non-overlapping:

	8 Feb. (n=10)	14 Feb. (n=2)
Lipid index (%)	18.5-26.6	5.6-9.8
Muscle index (SMV)	.281-.356	.217-.272

All birds in the 8 February sample (i.e. at the start of the severe

Table 1. Nutritional Condition of Three Species of Shorebirds in North-East England at the Beginning and End of a Period of Severe Weather in February 1978.

Species	Age	Condition Indices	Beginning			End			P (Student's t)
			Date	n	x̄ ± 1 s.e.	Date	n	x̄ ± 1 s.e.	
Dunlin	Ad.	Lipid Index (%)[a] (L.I.)	9 Feb	15	9.70 ± 0.66	15 Feb	4	9.90 ± 0.30	n.s.
		Muscle Index[b] (S.M.V.)		15	.281 ± .005		4	.272 ± .012	n.s.
		Muscle Index (%)[c] (L.D.M.)		15	5.75 ± 0.07		4	5.88 ± 0.16	n.s.
		Lean Weight (gm.) (L.W.)		15	48.30 ± 0.95		4	47.22 ± 0.65	n.s.
Dunlin	Juv.	L.I.	9 Feb	11	8.49 ± 0.61	15 Feb	6	7.75 ± 1.38	n.s.
		S.M.V.		11	.281 ± .007		6	.279 ± .011	n.s.
		L.D.M.		11	5.77 ± 0.10		6	5.98 ± 0.10	n.s.
		I.W.		11	47.59 ± 1.10		6	46.58 ± 1.06	n.s.
Bar-tailed Godwit[d]	Ad.	L.I.	9 Feb	3	11.30 ± 3.86	15 Feb	2	8.65 ± 2.01	n.s.
		S.M.V.		3	.278 ± .024		2	.255 ± .027	n.s.
		L.D.M.		3	6.19 ± 0.05		2	5.96 ± 0.46	n.s.
Golden Plover	Ad.	L.I.	8 Feb	10	24.16 ± 0.77	14 Feb	2	7.72 ± 2.07	<.001
		S.M.V.		10	.323 ± .008		2	.244 ± .023	<.01
		L.D.M.		10	7.75 ± 0.14		2	6.45 ± 0.48	<.05
		L.W.		10	163.64 ± 5.42		2	163.64 ± 5.13	<.05

a Lipid Index = $\frac{\text{fat (gm.)}}{\text{total body wt. (gm.)}} \times 100$

b Muscle Index (S.M.V.) = $\frac{\text{lean dry pectoral muscle (gm.)}}{\text{standard muscle volume}}$
(for details of method see Evans & Smith 1975)

c Muscle Index (L.D.M.) = $\frac{\text{lean dry pectoral muscle (gm.)}}{\text{lean weight (gm.)}} \times 200$

d Lean weights are omitted because of the large variations in body size in this species (Evans & Smith op. cit.).

weather) were carrying very similar lipid levels. Those Golden
Plover that remained on Lindisfarne moved to the intertidal mudflats
where they were subjected to inter-specific aggression, particularly
from Grey Plover (pers. obs., D.J. Townshend pers. comm.), as well
as encountering the same severe weather conditions as intertidal
shorebirds.

January - March 1979

The 1978/79 winter was the most severe for 16 years. The worst
conditions occurred between late December 1978 - mid January 1979,
and again in mid February 1979. No carcasses were analysed, but
sufficient live samples of Dunlin and Grey Plover were netted at
Teesmouth to make estimates of lipid indices, using formulae in
Davidson (unpublished).

1.   Grey Plover

Dugan et al. (in press) document the normal seasonal changes
in total body weight and those in winter 1978/79. The only sample
with weights significantly below normal was netted on 12 January,
when six birds averaged 137 gm. Since normal lean weights in winter
average 210 gm. (Davidson,unpublished), these birds must have metabolised all their available lipid reserves and they had, in addition,
lost up to 35% of their normal lean weight. Despite this extensive
use of lipids and muscle protein, most of these birds are known to
have survived, and to have regained their normal total body weights
by the end of February.

Lipid index estimates, calculated from total live weights,
(Figure 2), suggest that the Teesmouth population of Grey Plover
had not regained all their lipid reserves by the end of January
(i.e. two weeks later). However, this apparently low lipid index
could also have been obtained from the low live weight if lipid
levels had returned to normal, but lean weight was still below
normal (i.e. if the lean weight calculated from the formula was an
overestimate). It is not clear which of these two alternatives
occurred. Captive Mallard, Anas platyrhynchos, that had been starved,
regained 80% of their total weight within a week of unrestricted
food being made available (Jordan, 1953). Dugan et al. (in press)
attributed the poor nutritional condition of Grey Plover on 12
January primarily to periods of gales, and documented that this
species cannot feed successfully on open mudflats when the windspeed exceeds 25 knots.

2.   Dunlin

Lipid indices estimated from live birds caught during January
and February 1979 were consistently lower than during the less severe
weather of January and February 1980 (Figure 3). There were two

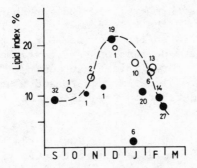

Figure 2. Lipid indices estimated from total body weights of Grey
Plovers at Teesmouth, during the 1978/79 winter (●), and other
years (○). Numbers give sample sizes, trend line fitted by eye.

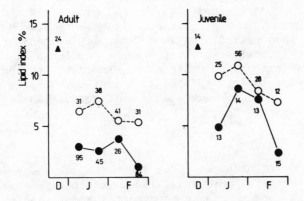

Figure 3. Lipid indices estimated from the total body weights of
adult and juvenile Dunlin Calidris a. alpina at Teesmouth during
the severe weather in Jan/Feb. 1979 (—●—) and during a mild
winter (Jan/Feb. 1980) (-- ○ --). Late December 1975 values (▲)
are shown for comparison. Numbers give sample sizes.

periods in early 1979 when lipid reserves appear to have been used
extensively: early January and late February. These coincided with
the two periods of most severe weather. In late December 1978 and
early January 1979 windchill was consistently high, with consecutive
days when the windchill factor was above 85. Conditions improved
in late January and early February, and Figure 3 shows that by early
February both adult and juvenile Dunlin had regained some of the
lipid reserves that had been previously mobilised. During mid
February, the second period of marked lipid index decline, there
was a period when windchill was above 100 on five consecutive days.

Juvenile Dunlin had higher lipid indices than adults during
January and February in the mild conditions of 1980 (Figure 3), but
the lipid indices of juveniles declined more rapidly during this
period. During the two periods of severe weather in early 1979,
juveniles metabolised more lipids than adults: in early January,
lipid indices were 3.5% lower in adults and 5% in juveniles in 1979
than in 1980; in mid February the differences were 4.3% in adults
and 5.1% in juveniles.

The low calculated lipid indices during the two periods of severe
weather in early 1979 could be underestimates if loss of lean weight
rather than lipids had occurred. However, for reasons discussed by
Evans and Smith (1975), if muscle protein was being metabolised during
these periods of severe weather, then lipid reserves were undoubtedly
also being used. The conclusion that nutritional reserves were
being used, therefore, remains valid, but the declines in lipid
reserves may be overestimates. Another possible explanation of the
low lipid levels in early 1979 could be that neither adult nor
juvenile Dunlin had attained their normal peak lipid indices (about
13%) in late December 1978. Unfortunately, very few Dunlin were
caught during this period, but an average lipid index of over 7%
in a sample of juveniles in early November conforms to the normal
lipid index pattern shown in Figure 1. Adherence to the normal
pattern by adults in late 1978 is more difficult to demonstrate
since few birds known to have stayed at Teesmouth throughout the
autumn were caught in December.

3.   Redshank

Samples of Redshank caught at Teesmouth in recent years are
too small for average lipid levels to be estimated so nutritional
changes in this species cannot be directly compared with those in
Dunlin and Grey Plover over the same period. However, even in mild
winters in north-east England (1971-73) both lipid levels (t_6 = 6.83
P < .001) and pectoral muscle size (t_5 = 3.63 P < .02) declined
significantly between November and late February (Figure 4). By late
February the lipid index had declined to only 2% compared with 7-8%
in Bar-tailed Godwit, Dunlin and Grey Plover (Figure 1). This decline in the condition of Redshank is comparable with that of Dunlin
during severe weather (Figure 3), but started before the mid-winter
peak of fat occurred in most shorebirds.

The highest incidence of gales in north-east England is during
November, December and January (Dugan et al., in press), so the decline in nutritional condition beginning in November would be consistent with any feeding difficulty caused by gales (further evidence
for this is examined later). It is surprising that, unlike plovers,
Redshank do not increase their lipid reserves more rapidly during
autumn to carry a higher lipid index than the 12-13% recorded in
November (Figure 4). However, during mild conditions, Redshank feed

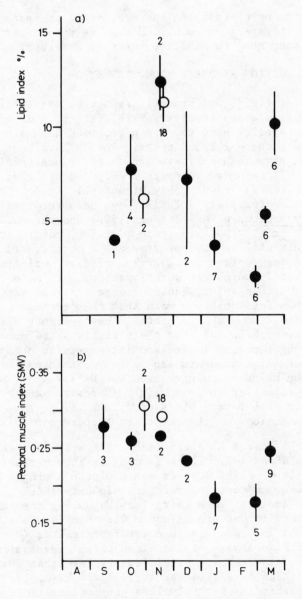

Figure 4. a). Lipid indices, and b). Pectoral muscle indices (in relation to Standard Muscle Volume) of Redshank in north-east England during 1971/72 and 1972/73 (●) and north-east Scotland during 1978 (○). Vertical bars are ± 1 standard error, and numbers give sample sizes.

for a greater part of the tidal cycle than most other intertidal shorebirds (Goss-Custard,1969; Pienkowski, 1973) so it is possible

that Redshank in north-east England are unable to increase their
rate of lipid storage in autumn, and that as soon as periods of high
winds begin, they need to mobilise nutrient reserves.

NUTRITIONAL CONDITION AT DEATH FROM STARVATION

Reserves of lipids and protein are not always sufficient to
ensure survival in exceptionally severe weather, and individuals of
many shorebird species have then been recorded as having starved to
death. In cases where fat and protein levels have been analysed,
lipid reserves have been exhausted at death, since the lipid indices
of less than 1.5% recorded (Marcstrom and Mascher 1979; Baillie and
Davidson,unpublished) are probably structural lipids. In passerines,
very similar structural lipid levels have been recorded in the trop-
ical bulbul Pycnonotus goiavier (Ward,1969) and the Bullfinch
Pyrrhula pyrrhula (Newton,1969). In addition, muscle protein has
been extensively utilised by the time death occurs, resulting in a
marked drop in lean weight. Comparison of death weights with
'normal' total body weight is unsatisfactory, as indicated by
Marcstrom and Mascher 1979), because large seasonal variations in
the size of the lipid reserves mean that the 'normal' weight also
changes markedly from month to month. (Lean weights also vary seasonally (Davidson,unpublished), but the variations are much smaller).
In the following section, inter- and intraspecific comparisons of
nutritional condition at death are made by examining losses only in
lean weight (below normal winter lean weight levels). Water content
decreases in proportion to lean weight (Marcstrom and Mascher 1979)
and so, although about two-thirds of a lean weight loss is a loss of
water, the conclusions on the use of protein remain valid.

Lean weight losses at death (Table 2) varied between 21% and
42%, averaging about 30%. There is considerable variation in the
extent of the lean weight loss at death within a single species.
There is, however, a constant relationship between the lean weight
level at death and body size (wing-length) during the same period
of severe weather in one area (Figure 5). Similar relationships
exist for normal lean weights in coastal shorebirds (Davidson,un-
published, and Figure 5), but the slope of the regression for birds
that had starved to death is significantly less that that for birds
in normal condition during winter (t_{11} = 3.14 P < .01). Small
species of shorebirds apparently lose proportionally less lean
weight at death than large species, possibly because small species
have higher metabolic rates per unit weight (Evans and Smith, 1975).
Shorebirds that do not use intertidal feeding areas extensively
during the non-breeding season do not conform to the same relationship of lean weight at death and body size.

Death from starvation during severe weather occurs most frequently in winter, but can also occur after shorebirds have arrived
on their breeding grounds in spring, during periods of heavy snowfall and low temperatures (Vepsalainen, 1968; Morrison, 1975; Marcstrom

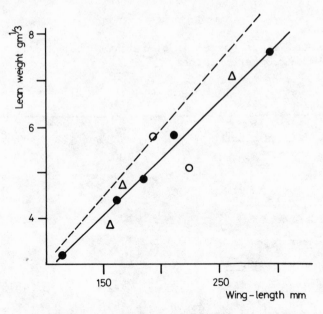

Figure 5. The relationship between wing-length and lean weight
in shorebirds that had starved to death during severe weather.
Solid circles are coastal species, open circles are inland species,
both from Ash (1964), and triangles are coastal species from other
sources. The regression (solid line), calculated for coastal
species from Ash (opt. cit.) only, is LW = .0247WL + .3881
r_4 = .9974 P < .001. The dashed line is the regression for adult
coastal shorebirds in normal winter condition in north-east England,
from Davidson (unpublished).

and Mascher, 1979). Because lean weights are higher in spring than in
winter (Davidson, unpublished), the absolute amount of lean weight lost
during these spring cold spells must be greater than that indicated
in Table 2.

     Differences in the percentage of lean weight lost at death are
likely to be correlated with the severity of the weather conditions.
When feeding is impossible, both lipids (as the main energy source)
and muscle protein (as a source of amino-acids) are metabolised
(Evans and Smith,1975). Once lipid reserves are exhausted, muscle
protein must provide the energetic requirements as well. However,
protein probably cannot be metabolised sufficiently fast when the
energetic requirements are very high. Thus birds may die at higher
lean weights during a brief, but very severe, period than during a
prolonged but less severe cold spell. In agreement with this, in
captive Mallard, Jordan (1953) recorded the smallest total weight
loss at death in birds starved during the winter, and the largest
in summer, despite the fact that Mallard carry more fat in winter
than summer (Owen and Cook, 1977).

Table 2. Lean Weight Losses in Shorebirds that had Starved to Death
During Severe Weather.

	Source[a]	Wt. at death[b]	Normal winter wt.[c]	% loss[d]
Redshank Tringa totanus	1	84.4	130.8	35.5
	2	106.1*	139.0	23.7
Oystercatcher	2	357.4*	467.4	23.5
Haematopus ostralegus	3	304.0*	385.8	21.2
Black-tailed Godwit	1	194.4	247.8	21.6
Limosa limosa				
Curlew Numenius arquata	1	439.9	663.3	33.7
Dunlin Calidris alpina	1	32.6	46.9	30.5
Turnstone Arenaria interpres	4	58.8	101.3	41.9
Golden Plover	1	113.8	170.8	33.4
Pluvialis apricaria	5	112.9		(33.9)
Lapwing Vanellus vanellus	3	137.9*	201.3	31.5
	1	132.4		(34.3)
	5	132.9		(34.0)
	6	134.1		(33.4)
	7	121.1		(39.9)
Woodcock Scolopax rusticola	1	193.8	?	?

a.   1 Ash (1964) Poole Harbour, Sussex, Jan/Feb. 1963; 2 Baillie
and Davidson (in prep.), Ythan, N.E. Scotland, Jan/Feb. 1979;
3 Marcstrom and Mascher (1979) South Sweden, April 1966; 4 Morrison
(1975) Ellesmere Is., Canada, June 1974; 5 Harris (1962) Skomer Is.,
Wales, Jan 1962; 6 Vepsalainen (1968) Finland, Spring 1966; 7 Creutz
and Piechocki (1970) E. Germany.
b.   Published lean weights are indicated by *. Other lean weights
were estimated by total body weight - 1% lipids.
c.   Except for Redshank from Baillie and Davidson (in prep.),
weights were calculated from Davidson (unpublished).
d.   Values in parentheses were calculated from normal winter lean
weights from different sources.

The shorebirds that are most susceptible to death by starvation fall into two categories: firstly, inland species, particularly Lapwing (Harris, 1962; Dobinson and Richards, 1964; Marcstrom and Mascher, 1979) and, secondly, three coastal species, Redshank, Oyster-catcher and Dunlin (Dobinson and Richards, 1964; Marcstrom and Mascher, 1979; Baillie, 1980). High mortality of Redshank is to be expected in view of the decline in their nutritional condition during even mild winters (see above). High mortality of Oystercatchers may be largely of juveniles (Baillie, 1980) whose weights decline rapidly between December and April on both the east and west coasts of Britain, whereas adults, unusually, gain weight gradually throughout the winter (Dare, 1977; Branson and Minton, 1978). No carcass analyses have been made on Oystercatchers, so the existence of a decline in pectoral muscle size in juveniles during mild winters, similar to that of Redshank, cannot be investigated. Possible reasons for mortality in Dunlin are discussed later.

RECOVERY FROM SEVERE LOSS OF CONDITION

Lipids can be mobilised and replaced rapidly, and temporary loss of all reserve lipids probably does not affect subsequent shore-bird survival, since Redshank lose almost all their lipid reserves by the end of February even in mild winters (Figure 4). Lipids can also become severely depleted during long non-stop migratory flights (e.g. Dick and Pienkowski, 1979). However, after depletion on migra-tion or during severe weather (Dugan et al., in press) they can be replaced rapidly. Additional lipid reserves can also be accumulated rapidly in advance of migration (e.g. Pienkowski et al., 1979; Summers and Waltner, 1979).

Reductions in muscle protein weights may occur during migration as well as winter starvation (Evans and Smith, 1975; Dick and Pienkowski, 1979). Depletions during winter (examined as lean weight) that did not result in death are listed in Table 3. Losses by both Redshank during mild winters, and Golden Plover during the severe weather in February 1978 are much less than the losses at which death has been recorded (21-42%). In contrast, the loss of 35% lean weight by Grey Plover during January 1979 is more than the losses at death under many circumstances, although, because the pectoral muscles are larger in plovers than in other shorebirds, plovers can probably utilise a greater percentage of their lean weight before dying. The level of lean weight from which recovery is possible must be very close to that at which death occurs, but this minimum recovery weight varies depending on species and severity of weather.

DISCUSSION AND CONCLUSIONS

Evidence is accumulating that the weather conditions that make it difficult for birds to maintain an energy balance during winter

Table 3. Lean Weight Losses Recorded in Live Shorebirds in North-
          East England During Severe Winter Weather

	Date	Lean wt. during severe weather (g.)	Normal winter lean wt. (g.)	% loss
Redshank Tringa totanus	Jan/Feb 1972/1973	117.2	139.0	15.6
Golden Plover Pluvialis apricaria	Feb 1978	163.6	180.4	9.3
Grey Plover P.squatarola	Jan 1979	135.6*	209.8	35.4

* estimated from total body weight - 1% lipids. Total body weight
from Dugan et al. (1981).

differ between shorebird species, even in mixed-species aggregations
feeding in the same habitat and area. During the 1978/79 winter at
Teesmouth, Grey Plover lost nutritional condition during cold weather
only when winds exceeding 25 knots prevented successful feeding on
open mudflats (Dugan et al., in press). Redshank wintering at Teesmouth during 1978/79 moved from the main mudflats to sheltered adjacent feeding sites during gales. At windspeeds greater than 25
knots, an average of 89% of the Tees population fed on sheltered
sites during low water, compared with only 30.5% when winds averaged
20-25 knots. Numbers of Redshank that stayed on the open mudflats
rather than moving to sheltered sites, were significantly lower on
days when the average windspeed exceeded 25 knots than on less windy
days (less than 25 knots $\bar{x}$ = 272 birds, greater than 25 knots
$\bar{x}$ = 46 birds, t_{11} = 3.62 P < .01). Some Curlew wintering on the
Tees estuary also moved from open mudflats during gales to feed on
the adjoining pastures (Townshend, this volume). In contrast, few
Dunlin chose to move from open mudflats to peripheral feeding areas
under any weather conditions during low water, even during gales
(only 7.5% of the Dunlin population moved even when winds exceeded
25 knots). However, during high water the percentage of the population that moved to peripheral sites and continued to feed after the
main mudflats had been covered by the rising tide significantly
increased with increasing windchill (r_{12} = 0.82 P < .001). An
average of 96% of the Tees Dunlin fed over the high water period
during days of high (>70) windchill in midwinter.

     Shorebirds may, therefore, be categorised according to their
responses to severe weather, as follows:

1.   Inland species (Lapwing, Golden Plover, probably Snipe) are the
     shorebirds most rapidly affected after the onset of severe
weather, since their feeding grounds become untenable (through freezing or snowcover) more rapidly and more frequently than those of
intertidal species. Gales may also limit feeding by Lapwing and
Golden Plover because they feed visually and buffeting may prevent
accurate prey location. These species move either to nearby coastal
habitats, or (particularly Lapwing) south or south-west to west
Britain, Ireland and exceptionally Iberia (Dobinson and Richards, 1964)
and Morocco (Smith, 1965). These are the only shorebirds for which
there is definite evidence of movements in direct response to severe
weather (Evans, 1967). Once Lapwing and Golden Plover have moved to
coastal areas they face the same problems during severe weather as
coastal plovers (see below) but, in addition, are attempting to feed
in a suboptimal habitat. Golden Plover have much higher lipid reserves and have larger pectoral muscles than other shorebirds in winter,
probably as an adaptation to these additional difficulties of winter
survival. There is no information on the normal nutritional condition of Lapwings in winter.

2.   Long-legged, and/or visual feeding estuarine species (e.g. Grey
     Plover, Redshank, Bar-tailed Godwit, Curlew, probably Ringed
Plover) have difficulty in foraging successfully on open mudflats
during gales, due to the direct physical effects of buffeting by the
wind. Plovers, because they feed only visually, may be more seriously
affected than other shorebirds in this category and they carry larger
lipid and protein reserves than the others. Bar-tailed Godwit
(Davidson unpublished) and Redshank (Goss-Custard, 1976) change from
visual to tactile feeding and move to feed in more sheltered areas,
but Bar-tailed Godwit may also cease feeding in severe gales even in
early autumn (Smith, 1975).

3.   Short-legged tactile feeding estuarine species (e.g. Dunlin,
     Knot) may be seriously affected only when very low temperatures
cause freezing of the mudflats, making feeding impossible, as in early
1963 (Dobinson and Richards, 1964; Pilcher, 1964). These species,
unlike Redshank (Goss-Custard, 1969) and Bar-tailed Godwit (Smith,
1975), cannot feed at the tide-edge, except in very shallow water.
They are not as directly affected by gales as the long-legged species:
buffeting is less severe, and tactile feeding is frequently employed.
Outside periods of very low temperatures, periods of high windchill,
causing increased energy requirements, are most important in inducing
nutrient reserve mobilisation.

4.   Open coastal species (e.g. Sanderling, Turnstone), particularly
     where they feed on sandy beaches, should require higher nutrient
reserves than other sandpipers because a). they frequently feed
visually and b). gales greatly reduce prey availability by causing
strong wave action that limits tide-edge feeding and disturbs the
substratum (Evans, 1976; Pienkowski, 1980a). Lack of shelter on sandy

beaches during gales and low temperatures may also cause increased
energy requirements. Limited information on the nutritional condition of Sanderling in north-east England (Davidson, 1980) indicates
that lipid reserves are maintained at higher levels than those of
Dunlin, Knot and Bar-tailed Godwit in late winter, but peak lipid
levels in these coastal species have yet to be investigated.

The greater use of nutritional reserves during severe winter
weather by plovers than by sandpipers correlates with the higher lipid
levels and larger pectoral muscles of plovers. This supports the
suggestion (Evans and Smith, 1975; Pienkowski et al., 1979) that winter
lipid levels are a trade-off between the need to carry sufficient
nutritional reserves for survival through periods of severe weather
and the disadvantages, probably energetic and aerodynamic, of carrying an increased weight-load of lipids during non-migratory periods
(Dick and Pienkowski, 1979; Pienkowski et al., 1979). Selection against
any increase in lipid levels above the minimum necessary could operate
through a reduced ability to avoid aerial predators: raptor predation
may also be an important factor in promoting winter flocking in shorebirds (Page and Whitacre, 1975; Smith, 1975). Since the nutritional
reserves carried after mid-winter can be inadequate to ensure survival
through severe weather, these opposing disadvantages of carrying
large nutritional reserves may be great. In Dunlin, juveniles carry
larger lipid reserves than adults during the winter, probably because
juveniles need to draw more extensively on their lipid reserves than
adults during severe weather. This phenomenon of higher lipid reserves in juveniles occurs also in Grey Plover and probably some other
shorebirds, and may be because some juveniles feed less successfully
than adults (Groves, 1978), particularly during adverse conditions
(Pienkowski, 1980b).

Short daylengths in mid-winter may be less important in determining the timing of peak lipid levels than supposed by Evans and
Smith (1975). Even in plovers, which feed visually (Pienkowski,
1980b), night feeding can be extensive. Indeed, in Lapwing and
Golden Plover feeding on pastures (Burton, 1974), food may be more
abundant at night than during daylight since earthworms are active
at the surface at night. Recent evidence (Pienkowski, 1980b; Dugan,
this volume) suggests that the coastal Grey Plover and Ringed Plover
also feed extensively (and successfully) at night.

Low temperatures per se may also be less important in determining peak lipid levels, and the use of nutritional reserves, than
supposed by Evans and Smith (1975) and Pienkowski et al. (1979).
However, low temperatures are important in depressing prey availability (Evans, 1979; Pienkowski, 1980a) which in turn may prevent
birds obtaining food at a sufficient rate to balance energy requirements. Mortality due to low temperatures alone may occur only when
intertidal areas freeze over during exceptionally severe winters
(Dobinson and Richards, 1964), thus preventing any feeding by shore-

birds. During most periods of severe weather it is the effects of
gales and high windchill, to which low temperatures of course contribute, that cause the most serious problems to the maintenance of
energy balance in shorebirds. Shorebirds can, at least partly,
counteract changes in prey availability induced by low temperatures
by changing their foraging behaviour (Smith,1975; Evans,1979;
Pienkowski,1980a,b; Davidson,unpublished), but they can avoid gales
and high windchill only by moving into shelter, which often reduces
the rate at which they can feed (Evans,1976; Dugan et al.,in press).

ACKNOWLEDGEMENTS

    I am grateful to Dr. P.R. Evans for much assistance and advice
during his supervision of this work, and to my colleagues P.J. Dugan,
L.R. Goodyer, Dr. M.W. Pienkowski and D.J. Townshend for help during
fieldwork and for making information readily available. Much of the
study was supported by a NERC Research Studentship, and work during
1978/79 was jointly funded by Cleveland County Council and the Nature
Conservancy Council.

REFERENCES

Ankney, C.D. 1977. The use of nutrient reserves by breeding Lesser
       Snow Geese Chen c.caerulsecens. Can. J. Zool. 55: 984-987.
Ash, J.S. 1964. Observations in Hampshire and Dorset during the
       1963 cold spell. Brit. Birds 57: 221-241.
Baillie, S. 1980. The effect of the hard weather of 1978/79 on the
       wader populations of the Ythan Estuary. Wader Study Group Bull.
       No. 28: 16-17.
Branson, N.J.B.A. and Minton, C.D.T. 1978. Wash Wader Ringing Group
       - Report 1975-76.
Burton, P.J.K. 1974. Feeding and the feeding apparatus in waders:
       a study of the anatomy and adaptations in the Charadrii.
       British Museum (Natural History). London. pp. 150.
Creutz, U. and Piechocki, R. 1970. Uber Durchschnitts- und Minimal-
       gewichte von Vogeln verschiedener taxonomischer Ordnungen.
       Falke 17: 42-47.
Dare, P.J. 1977. Seasonal changes in body-weight of Oystercatchers
       Haematopus ostralegus. Ibis 119: 494-506.
Davidson, N.C. 1979. Changes in the body composition of shorebirds
       during winter. Wader Study Group Bull. No. 26: 29-30.
Davidson, N.C. 1980. Winter nutritional condition of Sanderling in
       north-east England. Wader Study Group Bull. No. 30: 20-21.
Dick, W.J.A. and Pienkowski, M.W. 1979. Autumn and early winter
       weights of waders in north-west Africa. Ornis Scand. 10:
       117-123.
Dobinson, H.M. and Richards, A.J. 1964. The effects of the severe
       winter of 1962/63 on birds in Britain. Brit. Birds 57: 373-434.
Dugan, P.J., Evans, P.R., Goodyer, L.R. and Davidson, N.C. In press.
       Winter fat reserves in shorebirds: disturbance of regulated

levels by severe weather conditions. <u>Ibis</u>.

Evans, P.R. 1976. Energy balance and optimal foraging strategies
   in shorebirds: some implications for their distributions in
   the non-breeding season. <u>Ardea</u> 64: 117-139.

Evans, P.R. 1979. Adaptations shown by foraging shorebirds to
   cyclical variations in the activity and availability of their
   intertidal invertebrate prey. pp. 357-366 in 'Cyclic Phenomena
   in Marine Plants and Animals' ed. E. Naylor & R.G. Hartnoll.
   Pergamon Press. Oxford & New York.

Evans, P.R., Herdson, D.M., Knights, P.J. and Pienkowski, M.W. 1979.
   Short-term effects of reclamation of part of Seal Sands,
   Teesmouth, on wintering waders and Shelduck. <u>Oecologia</u> 41:
   183-206.

Evans, P.R. and Smith, P.C. 1975. Studies of shorebirds at Lindisfarne, Northumberland. II. Fat and pectoral muscles as indicators of body condition in the Bar-tailed Godwit. <u>Wildfowl</u> 26:
   64-76.

Fuller, R.J. and Youngman, R.E. 1979. The utilisation of farmland
   by Golden Plovers wintering in southern England. <u>Bird Study</u>
   26: 37-46.

Goss-Custard, J.D. 1969. The winter feeding ecology of the Redshank
   <u>Tringa</u> <u>totanus</u>. <u>Ibis</u> 111: 338-356.

Goss-Custard, 1976. Variations in the dispersion of Redshank <u>Tringa</u>
   <u>totanus</u> on their winter feeding grounds. <u>Ibis</u> 118: 257-263.

Groves, S. 1978. Age-related differences in Ruddy Turnstone foraging
   behaviour. <u>Auk</u> 95: 95-103.

Harris, M.P. 1962. Weights of five hundred birds found dead on
   Skomer Island in January 1962. <u>Brit. Birds</u> 55: 97-103.

Jordan, J.S. 1953. Effects of starvation on wild Mallards. <u>J.</u>
   <u>Wildl. Mgmt.</u> 17: 304-311.

Kendall, M.D., Ward, P. and Bacchus, S. 1973. A protein reserve in
   the pectoralis major flight muscle of <u>Quelea</u> <u>quelea</u>. <u>Ibis</u> 115:
   600-601.

Marcstrom, V. and Mascher, J.W. 1979. Weights and fat in Lapwings
   <u>Vanellus</u> <u>vanellus</u> and Oystercatchers <u>Haematopus</u> <u>ostralegus</u>
   starved to death during a cold spell in spring. <u>Ornis</u> <u>Scand</u>.
   10: 235-240.

Morrison, R.I.G. 1975. Migration and morphometrics of European
   Knot and Turnstone on Ellesmere Island, Canada. <u>Bird Banding</u>
   46: 290-301.

Newton, I. 1969. Winter fattening in the Bullfinch. <u>Physiol. Zool</u>.
   42: 96-107.

Owen, M. and Cook, W.A. 1977. Variations in body weight, wing-
   length and condition of Mallard <u>Anas</u> <u>platyrhynchos</u> <u>platyrhynchos</u>
   and their relationship to environmental changes. <u>J. Zool.,Lond</u>.
   183: 377-395.

Page, G. and Whitacre, D.F. 1975. Raptor predation on wintering
   shorebirds. <u>Condor</u> 77: 73-83.

Pienkowski, M.W. 1973. Feeding activities of wading birds and shel-
   ducks at Teesmouth and some possible effects of further loss of

habitat. Report to the Coastal Ecology Research Station (Nature
    Conservancy).

Pienkowski, M.W. 1980a. Differences in habitat requirements and
    distribution patterns of plovers and sandpipers as investigated
    by studies of feeding behaviour. Proc. Int. Waterfowl Research
    Bureau Symposium on Feeding Ecology. Gwatt, Switzerland, Sept.
    1977. Verh. orn. Ges. Bayern. 23: in press.

Pienkowski, M.W. 1980b. Aspects of the ecology and behaviour of
    Ringed and Grey Plovers Charadrius hiaticula and Pluvialis
    squatarola. Unpubl. Ph.D. Thesis, University of Durham.

Pienkowski, M.W., Lloyd, C.S. and Minton, C.D.T. 1979. Seasonal and
    migrational weight changes in Dunlins Calidris alpina. Bird
    Study 26: 134-148.

Pilcher, R.E.M. 1964. Effects of the cold weather of 1962-63 on
    birds of the north coast of the Wash. Wildfowl Trust Ann.
    Report 15: 23-26.

Smith, K.D. 1965. On the birds of Morocco. Ibis 107: 493-526.

Smith, P.C. 1975. A study of the winter feeding ecology and behaviour
    of the Bar-tailed Godwit (Limosa lapponica). Unpubl. Ph.D.
    Thesis, University of Durham.

Summers, R.W. and Waltner, M. 1979. Seasonal variations in the mass
    of waders in Southern Africa, with special reference to migra-
    tion. Ostrich 50: 21-37.

Vepsalainen, K. 1968. The effect of the cold spring 1966 upon the
    Lapwing Vanellus vanellus in Finland. Ornis Fennica 45: 33-47.

Ward, P. 1969. Seasonal and diurnal changes in the fat content of
    an equatorial bird. Physiol. Zool. 42: 85-95.

THE IMPORTANCE OF NOCTURNAL FORAGING IN SHOREBIRDS: A CONSEQUENCE OF INCREASED INVERTEBRATE PREY ACTIVITY

P.J. Dugan

Department of Zoology
University of Durham
South Road, Durham

INTRODUCTION

Many shorebirds are known to feed at night. However, the extent to which this contributes to the total daily energy intake of different individuals under different weather conditions and at different times of year is largely unknown. Several authors, using a variety of methods, have attempted to quantify nocturnal food intake of Oystercatchers, Haematopus ostralegus. Drinnan (1957) observed no difference between night and day in the number of shells opened and left on the mud, as did Davidson (1967) by direct observation, using infra-red equipment. However, Heppleston (1971) and Hulscher (1974) using captive birds, obtained night values of 0.58 and 0.86 respectively, of the intake during the day. In Knot, Calidris canutus, feeding on Macoma balthica, Prater (1972) found that the stomach contents of birds shot after daylight feeding periods contained four times the quantity of prey found in night feeding birds.

For several years most estimates of nocturnal food intake by shorebirds of other species have been extrapolated from these studies as, although the reliability of the results has not been examined in detail, they are consistent with the intuitive assumption that the foraging success of tactile predators should be relatively unaffected by darkness, but that of visual predators should be greatly impaired. The intake of nocturnally feeding, visually foraging, shorebirds such as plovers is normally quoted as approximately 50% of that taken during the day (Schramm, 1978). More recently, further studies using night viewing equipment have tended to confirm the assumptions about foraging success (Knights,1979; Pienkowski, 1980). Pienkowski compared the pacing and peck rates of plovers during daylight and darkness, yet because during darkness, prey items cannot always be recognised, the relationships between these measures and rates of energy or biomass intake are unknown. It is assumed that they remain the same at night as during the day and, consequently, that the lower nocturnal peck rate indicates a lower energy intake rate. However, in making this assumption the activity of the various invertebrate prey, of crucial importance to the foraging behaviour of plovers, (Pienkowski,1980), has normally been ignored. It is well known that terrestrial annelids, particularly Lumbricus terrestris are nocturnally active. Further, it has been shown that shorebirds feeding on this species are more active and more successful by night than by day (MacLennon,1979). However, no comparable information exists on the nocturnal activity of the intertidal prey of shorebirds. Furthermore, it has been assumed that the absence at night of shorebirds from their daylight feeding grounds implies that the birds are not feeding. The alternative possibility, that individuals change their feeding location at night has rarely been considered.

This paper describes an investigation, through study of the behaviour of both predator and prey, of the importance of nocturnal feeding in Grey Plover Pluvialis squatarola wintering on the Tees estuary in N.E. England.

METHODS

Feeding Observations

As part of a study of the behavioural ecology and population dynamics of wintering Grey Plover on the Tees estuary, rates of prey intake by day were measured by direct observation by telescope. Birds feeding on the main intertidal mudflats were observed at close range (10-50 metres) from a hide mounted on a rubber dinghy and beached on the mud on the ebbing tide. The full low water period was monitored. In mid winter, on the days for which data are presented, this spanned almost all of the daylight hours.

In the study area, there is a limited variety of macrofauna. Only 4 species are common, Nereis diversicolor; Hydrobia ulvae; Macoma balthica and Corophium volutator, of which only the first two are abundant (Evans et al., 1979). Consequently, each prey taken was normally identified with confidence. The sizes of Nereis taken were estimated by comparison with the bill lengths of the Grey Plovers ($\bar{x}$ = 29.0 mm. Prater et al., 1977), a method which has been shown to be accurate, to within 15% by weight (Dugan,1981). Calorific values of the prey were determined for different periods of the year using a Phillipson Microbomb Calorimeter (Phillipson, 1964).

Prey Availability

Invertebrate activity was measured within a 50cm. x 50cm. quadrat laid on the mud surface. This area was scanned intensively for 2 minutes and all animals active at the surface of the mud recorded. For Nereis virens the number of defaecatory heaps in each quadrat was counted. At night a torch was used to illuminate the surface of the mud. This had no visible effect on animals already active on the surface.

The depth and temperature of the water at the surface of the mud, and air and mud (at 2 cm. depth) temperature were recorded at the end of each observation period.

Location of Feeding Sites

Two adult Grey Plovers were caught with cannon nets, individually colour-ringed, instrumented with AVM L-module transmitters as back-packs, kept overnight for observation and then released. Transmitter and harness weighed an average of 20.0 grams or 8.0% of the average birds' weight. Bird locations were monitored from the ground with an AVM LA-12 receiver and hand held antenna. Three or more cross bearings using a prismatic compass were obtained for each bird.

RESULTS

Energy Intake

In both mild and severe winter weather conditions, the calorific intake by day of flock-feeding Grey Plovers at Teesmouth was rarely above that required to meet their basal metabolic requirements of 25 Kcals/day^{-1}, estimated from the equation of Lasiewski and Dawson (1967) (Table 1). Despite the difficulties in measuring intake it is unlikely that the values obtained in this study are underestimates. Not only were birds observed at very close range for the whole day-light feeding period and the method of prey size estimation shown to be accurate (Dugan, 1981), but as reported elsewhere (Dugan, 1981) using the same procedures, values in excess of 100 cals/min^{-1} were obtained consistently for territorial (as opposed to flocking) birds feeding in other areas of the mudflats.

The most conservative estimate of the energy requirements of a free-living shorebirds in mild conditions is 3 x B.M.R. (Evans et al., 1979). This is far above the estimates of daylight intake reported here. Supporting evidence for the reality of the low food intake rates of Grey Plovers by day come also from other studies. Pienkowski (1980) obtained values as low as 0.53 and 0.83 x B.M.R. for Grey Plover at Lindisfarne, N.E. England and Schramm (1978) calculated a value of only 1.18 x B.M.R. for Grey Plover in South Africa.

Table 1.   Diurnal Energy Intake of Flock Feeding Birds

Date	n	x (cals min^{-1})	S.E.	Total intake during L.W. period (Kcals)	xB.M.R.
09.10.78	10	36.9	7.4	17.7	0.71
21.10.78	10	70.2	7.1	33.7	1.35
22.10.78	9	51.4	12.4	24.7	0.99
20.11.78	7	47.4	14.0	22.8	0.91
21.12.78	7	15.1	2.7	7.3	0.29
03.01.79	11	44.2	9.0	21.2	0.85
04.01.79	8	45.2	9.7	21.7	0.87
15.01.79	10	57.7	21.6	27.7	1.11
04.02.79	8	37.6	9.5	18.1	0.72
20.08.79	4	139.7	24.0	67.1	2.68
07.09.79	7	68.9	21.0	33.1	1.32
08.01.80	6	75.8	14.3	36.4	1.46

Intake values refer to the mean rate among all individually recognisable birds observed during the low water period on each date. Rates were calculated for total observation time i.e. feeding time plus time spent engaged in other activities e.g. preening, aggression. Only data from birds watched for three, or more, minutes were used. n = number of birds watched. The length of the L.W. period i.e. that for which the feeding grounds are exposed on each tidal cycle, is approximately eight hours. B.M.R. was calculated as 25Kcal day^{-1} using the formula for non-passerines given by Lasiewski and Dawson (1967). Lean weight of wintering Grey Plovers = 209.8 ± 5.7 grams n = 5 (Davidson, this volume).

Nocturnal Activity of Prey and Feeding Rates of Plovers

Pienkowski (1980) observed Grey Plovers feeding at night and found that their peck rates remained the same as, or fell below, those during the day. He concluded that this almost certainly implied a reduced rate of biomass intake. However, if prey behaviour changes at night in such a way as to increase the availability of large items, the predator may peck less often per minute but obtain a greater biomass at each peck than during daylight. The nocturnal observations of prey activity reported in Table 2 show that an increase in availability does occur in the species studied. During daylight, no Nereis diversicolor could be detected at the mud sur- face, either during observations of the quadrats nor during many other visits to the same mudflats. In contrast, although the numbers seen varied greatly from night to night in a manner unrelated to obvious environmental variables (notably temperature), Nereis were detected at the surface on every occasion at night, the anterior end protruding from the burrow in a manner similar to that described by Cram and Evans (1980). During the quadrat observations only small Nereis, up to c. 4 cm. in length, were seen within the selected areas, although larger individuals, up to c. 10 cm. in length, were observed when surveying larger areas of the mud surface close by. It is likely that the absence of large Nereis from the quadrats arose from the relatively low density and patchy occurrence of large animals. Further, if the marked nocturnal increase in activity of worms at the surface, detectable by a human observer, is also detect- able by Grey Plovers, then a much higher proportion of all Nereis

Table 2. Surface Activity of Invertebrate Prey of Grey Plover on Seal Sands

Date	Nereis x ± S.E.		Hydrobia x ± S.E.		Corophium x ± S.E.		Temperature (^{o}C) Air	Water	Mud
Daylight									
10.12.79	0.0	0.0	3.0	1.7	0.0	0.0	5.0	7.0	6.0
12.12.79	0.0	0.0	-	-	-	-	8.0	6.3	7.0
Darkness									
09.12.79	7.0	1.7	40.0	8.8	1.3	1.0	8.0	6.0	7.0
19.01.80	4.0	1.4	10.0	3.6	1.0	0.7	3.5	4.0	4.0
18.02.80	2.0	1.0	15.0	1.7	2.0	1.0	6.5	4.5	5.5
02.03.80	2.0	1.7	3.0	0.7	3.0	1.7	4.0	4.0	4.0
05.03.80	1.0	0.7	6.0	3.0	0.0	0.0	5.0	4.5	5.0
10.03.80	2.0	1.0	13.0	4.6	1.0	0.7	1.5	3.5	3.5

All observations were made one hour after exposure of the study site. Values represent mean densities per square metre calculated from the number of animals observed at the surface of each 0.25 metre square quadrat. On all dates n=4.

present, including large individuals, will be available for capture at any one moment at night. It is crucial to the hypothesis put forward here that larger individuals can be detected and are taken by the birds.

The increase in availability of Nereis by night is, I suggest, sufficient to offset any decrease due to poorer vision during darkness. Indeed, while such impairment almost certainly occurs, it is possible that this may not be as severe for plovers as for other shorebirds. The relatively large eye, in relation to head size, of plovers, compared to that of sandpipers, is normally assumed to be an adaptation to enhance visual detection and capture of almost all prey at the mud surface by day. It is probable that the large eye is also an advantage to plovers while feeding at night. Indeed, it is thought that this group of shorebirds evolved in arid environments where most prey are active only by night (Pienkowski 1980).

Locations of Nocturnal Foraging by Grey Plovers

The activity of Nereis declines with time after the mud is uncovered, as the depth of water film on the surface decreases. By three hours after first exposure, very few Nereis are detectable at the surface (Fig. 1). If birds remained to forage in such areas their rate of biomass intake would decline to below that during the day. However, the results of the telemetry study suggest that birds feeding in the flock by day change their feeding site at night. One individual (Fig. 2, Bird A), which fed during most days on an area of firm mud, spent 3 of the 5 observation nights on the other side of the estuary on very soft mud where no plovers foraged by day. Of the other two nights studied, one was of extreme spring tides, and the bird fed at the lowest tidal levels on Seal Sands, not mormally uncovered at low water. The other night was of extreme neap tides and the bird again fed on Seal Sands, but at mid tidal levels, probably because both the areas used on the other four nights remained submerged.

The feeding location of the other radio-tagged Grey Plover could not be monitored so precisely because of poor directionality of the signal at close range, and the habit of the bird of feeding very close to the road from which the signals were monitored. Despite this limitation the bird was discovered feeding consistently during the day on an area of soft mud on Seal Sands, and detected on most nights within this same area or nearby, although it was not possible to determine whether the bird was feeding on exactly the same type of substratum as by day. On the one night of extreme spring tides, however, when much lower inter-tidal areas were exposed, it moved 2km to another feeding site on a marine beach (Fig. 2, Bird B).

No observations of prey activity were made at the nocturnal feeding sites of these radio-tagged birds on Seal Sands but it is

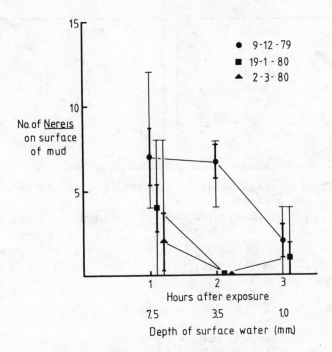

Figure 1. Surface activity of Nereis diversicolor in relation to
time after exposure of, and depth of surface water on, the study
site. Thick and thin vertical lines represent one standard error
and maxima and minima.

likely that both birds were feeding on areas of wet mud where Nereis
remained active throughout the low water period. On one area of wet
mud examined, Nereis were found at the surface after low water when
none were detectable on the surface of the surrounding mud from which
the water film had disappeared.

     Observation on the south side of the estuary, at the regular
feeding site of bird A, revealed that in this area (and also on the
marine beach used on one occasion by bird B) the much larger Nereis
virens replaces Nereis diversicolor as the dominant polychaete.
This species also displayed a difference in activity by day and by
night (Table 3). While the data are more limited, it is clear that
on some nights individual worms defaecating at the surface should
have been detectable to birds, just as Arenicola marina is detectable
during the day (Smith,1975; Pienkowski,1980). (The night on which
no increase in activity was detected was cold and Smith (1975) has
shown that diurnal activity of Arenicola is also depressed by such
temperatures). In view of the large size of Nereis virens, the use
of these nocturnal foraging sites by the birds may well have provided

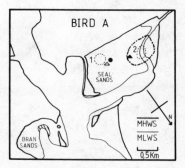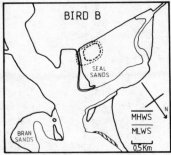

Figure 2. Feeding areas used, during daylight and darkness, by
transmitter bearing birds.

Bird A. Daylight: (—·——·) 17-2-80;
11-2-80; 23-2-80; 24-2-80; 7-3-80.
All times L.W.

Bird B. Daylight (—·——·)
17-2-80; 18-2-80; 23-2-80;
7-3-80. All times L.W.

Darkness \\\\\\\ 25-2-80 L.W.
                       2-3-80 L.W.
                       5-3-80 L.W.
      1. ········ 18-2-80 L.W. - 1.5hr
                       L.W. + 1.5hr
      2. ········ 18-2-80 L.W. + 1.0hr
            ▲ 10-3-80 L.W. - 1.5hr
            △ 10-3-80 L.W.
            ○ 2-3-80 L.W. - 1.5hr

Darkness \\\\\\ 18-2-80 L.W.
      ----- 25-2-80 L.W.
      ------ 2-3-80 L.W.
      ------ 5-3-80 L.W.
               - 1.5hr

them with a very high rate of biomass intake, considerably higher
than that possible during the day when the fed only on _Nereis
diversicolor_.

Table 3. Surface Activity of _Nereis virens_ on Bran Sands.

Date	No. of Casts x ± S.E.		Temperature (°C)		
			Air	Water	Mud
Daylight					
21-3-80	2.0	1.0	2.0	5.0	5.0
23-3-80	4.0	1.0	5.5	5.5	6.0
Darkness					
5-3-80	42.0	7.7	5.0	4.5	5.0
23-3-80	4.0	1.0	0.5	2.5	2.0

All observations were made between L.W. and L.W. +2. Values represent mean densities per square metre calculated from the number of
defaecatory casts observed in each 0.25 metre square quadrat.
n = 4 on 5-3-80 and 8 on all other dates.

DISCUSSION

The evidence presented here suggests that, on the Tees estuary, the major part of the energy requirements of some individual Grey Plover are met at night rather than by day. It is likely that this phenomenon is widespread in the population. Not only does the number of flock-feeding birds in winter frequently exceed 150, but evidence reported elsewhere, (Dugan,1981), suggests that some territorial birds may not achieve sufficient intake during daylight. Nereis virens is abundant in several areas at Teesmouth in addition to those used by the two radio-tagged birds and it is likely that these (particularly at lower tidal levels on the northern edge of Seal Sands) serve as major nocturnal feeding grounds for Grey Plover and possibly other species of shorebirds.

The critical test of the hypothesis advanced here lies in detailed direct measurement of food intake of birds foraging at night. It is hoped that this will soon be undertaken, but until then great care must be taken in assessing the importance of daylight feeding in other species and other areas.

The present study resulted because energy intake during daylight was observed to be low. However, in studies of other species and in other areas where observed diurnal intake approximates to expected daily energy requirements, the data have been assumed to be complete and it has been concluded that requirements are met largely by diurnal intake. Clearly, the conclusions of the present study should be extended to other areas and species with caution. The relationship suggested here depends upon the complex behavioural interaction between foraging behaviour used by plovers and the activity and distribution of their major prey. All of these vary between shorebird species and the latter two vary greatly within the wintering range of each species. However, particularly in view of the estimated and untested nature of the values of metabolic rate for free-living birds, caution in interpretation of diurnal energy intake is required until similar studies can be made of other species and other areas and a much fuller understanding of the importance of nocturnal feeding is achieved.

ACKNOWLEDGEMENTS

I thank L.R. Goodyer for catching birds to be radio-tagged, D.J. Townshend for an introduction to telemetry and M.W. Pienkowski for discussion on nocturnal foraging in plovers. Dr. P.R. Evans read, and made valuable comments on a first draft of the paper. My study on the Tees was supported by an N.E.R.C. studentship. Radio-tagging of birds on the Tees forms part of a study financed by an S.R.C. research grant to P.R. Evans.

REFERENCES

Cram, A. and Evans, S.M. 1980. Stability and lability in the
    evolution of behaviour in Nereid polychaetes. Anim. Behav.
    28: 483-490.
Davidson, P.E. 1967. The Oystercatcher as a predator of commercial
    shellfisheries. Ibis 109: 473-474.
Dugan, P.J. 1981. Seasonal movements of shorebirds in relation to
    spacing behaviour and prey availability. Unpublished Ph.D.
    thesis, University of Durham.
Drinnan, R.E. 1957. The winter feeding of the Oystercatcher
    (Haematopus ostralegus) on the edible cockle (Cardium edule).
    J. Anim. Ecol. 26: 441-469.
Evans, P.R., Herdson, D.M., Knights, P.J. and Pienkowski, M.W. 1979.
    Short-term effects of reclamation of part of Seal Sands,
    Teesmouth, on wintering waders and Shelduck. Oecologia 14:
    183-206.
Heppleston, P.B. 1971. The feeding ecology of Oystercatchers
    (Haematopus ostralegus. L.) in winter in Northern Scotland.
    J. Anim. Ecol. 40: 651-672.
Hulscher, J.B. 1974. An experimental study of the food intake of
    the Oystercatcher Haematopus ostralegus L. in captivity during
    the summer. Ardea 62: 155-171.
Knights, P.J. 1979. Effects of changes of land use on some animal
    populations. Unpublished Ph.D. thesis, University of Durham.
Lasiewski, R.C. and Dawson, W.R. 1967. A re-examination of the
    relationship between standard metabolic rate and body weight
    in birds. Condor 69: 13-23.
MacLennon, J.A. 1979. Formation and function of mixed species wader
    flocks in fields. Unpublished Ph.D. thesis, University of
    Aberdeen.
Phillipson, J. 1964. A miniature bomb calorimeter for small biolo-
    gical samples. Oikos 15: 130-139.
Pienkowski, M.W. 1980. Aspects of the ecology and behaviour of
    Ringed and Grey Plovers, Charadrius hiaticula and Pluvialis
    squatarola. Unpublished Ph.D. thesis, University of Durham.
Prater, A.J. 1972. The ecology of Morecambe Bay. The food and
    feeding habits of Knot (Calidris canutus L.) in Morecambe Bay.
    J. Appl. Ecol. 9: 179-194.
Prater, A.J., Marchant, J.H. and Vuorinen, J. 1977. Guide to the
    identification and ageing of Holarctic waders. British Trust
    for Ornithology, Field Guide No. 17. B.T.O. Tring.
Schramm, M. 1978. The feeding ecology of Grey Plover on the Swartkops
    estuary. Unpublished B.Sc. thesis, University of Port Eliza-
    beth.
Smith, P.C. 1975. A study of the winter feeding ecology and behaviour
    of the Bar-tailed Godwit, Limosa lapponica. Unpublished Ph.D.
    thesis, University of Durham.

THE IMPORTANCE OF FIELD FEEDING TO THE SURVIVAL OF WINTERING MALE

AND FEMALE CURLEWS <u>NUMENIUS ARQUATA</u> ON THE TEES ESTUARY

D.J. Townshend

Department of Zoology
University of Durham
Science Laboratories
South Road, Durham, DH1 3LE

INTRODUCTION

Curlews, <u>Numenius arquata</u>, wintering on the Tees estuary use
two feeding habitats, the mudflats and the adjacent fields. In mid
winter, numbers feeding on intertidal areas fall, but numbers counted
on and around the estuary at high water are unchanged. This paper
describes the seasonal changes in use of the two habitats shown by
individual Curlews. The role of the weather in determining the
changes is considered. Differences in behaviour between male (short-
billed) and female (long-billed) Curlews are discussed in relation
to the depth distribution of their prey. These behavioural strategies are related to the different energy requirements of the sexes.

METHODS

Study area

The study was carried out on the Tees estuary, north-east
England (54° 37' N, 1° 12' W) during the two winters of 1976/77
and 1977/78. Additional information in the 1978/79 winter was
collected by N.C. Davidson, F.L. Symonds and L.R. Goodyer (unpublished data). The estuary comprises two large intertidal areas, Seal
Sands and Bran Sands. Seal Sands is the most productive area and
formed the main intertidal feeding area for Curlews.

Counts

The number of Curlews present on Seal Sands at low water was
counted, using a 15-60x60 telescope, at about fortnightly intervals,

beginning in October 1976. Counts for Bran Sands during the 1977/78
winter were supplied by D.M. Brearey (unpublished data). Regular
counts were also made on known roost sites at high water, when the
major intertidal feeding areas were covered. High water counts are
minima as small numbers of Curlews may have roosted or fed on the
fields undetected (see below). The counts made for the Birds of
Estuaries Enquiry are also minima as roosts may have been missed.
Therefore, only those high water counts that exceed the low water
counts for Seal Sands have been included in the analysis. Immediately to the north-west and west of the estuary lie extensive areas of
rough pasture. Although it was impossible to check all these fields
for feeding Curlews, counts were made at both low and high water at
frequent intervals on one easily accessible pasture used by the
Curlews, the Brinefields. This site was known to be favoured by
Curlews in previous years (Pienkowski, 1973; Knights, 1979).

Marked Birds

To permit long-term observations on individual birds, 82 Curlews
were uniquely marked with combinations of colour rings, mainly in
the spring and summer of 1977. During 1976/77 and 1977/78 checks
were made for the presence of colour-marked Curlews on Seal Sands
about every 4 days and on the Brinefields pasture at about weekly
intervals. As birds on the Brinefields were regularly disturbed by
brine extraction operations, and the major part of the study concerned
the feeding behaviour of birds on Seal Sands, sightings of colour-
marked Curlews on the fields were much less frequent than on the
mudflats. In 1978/79 checks for marked Curlews were made on the
pastures (N.C. Davidson, pers. comm.) giving the data the opposite
bias. The data are analysed with these biases in mind.

Curlews are sexually dimorphic, the females being larger (e.g.
Witherby et al. 1945; Bannerman, 1961). The most clear-cut difference between the sexes is in the length of the bill. As estimates
of the bill length separating males from females vary (Prater et al.,
1977; Elphick 1979; Bainbridge & Minton in prep.) the division used
in this study (males <122 mm bill length, females >129 mm) was based
upon the biomodal distribution of bill lengths of the adult colour-
marked birds (Townshend, 1981). Using this separation, 48 of the
colour-marked Curlews were males, 29 females and 5 were of intermediate bill length and could not, therefore, be sexed with confidence.

Foraging Behaviour

The main prey of Curlews on the mudflats was the polychaete
Nereis diversicolor, and on the fields, earthworms. On both
habitats Curlews ate other invertebrate prey only extremely rarely.
Measures of foraging by Curlews were made on the mudflats in order
to identify any differences between males and females. On 6 days
during the 1977/78 winter detailed and extensive observations were

made on both a marked male (Y W/R) and a marked female (R O/O) which
fed near to each other on the same mudbank on Seal Sands throughout
the winter (and were never seen on the fields). Additional observations on one or other of the birds were made on other occasions.
Because there was a marked difference in bill length between the two
birds (male 114mm, female 156mm) any differences in foraging abilities between short - and long-billed birds should be revealed by this
analysis.

During observation periods lasting 6-38 minutes I recorded the
number and length of all Nereis worms swallowed (whole or broken).
The length of each worm was estimated in relation to the bill length
of the bird extracting it, and later converted to an estimated length
(mm.). From this information an estimate of the rate of biomass
(dry weight) intake was made. Each length estimate of whole or
broken worms was converted to an estimate of dry weight using an
empirical relationship between length and body weight derived by
Ratcliffe (1979):

$$\log_{10} \text{ dry weight (mg.)} = 3.072 \log_{10} \text{ head width (mm.)} + 0.385,$$

$$\text{where head width (mm.)} = \frac{\text{body length (mm.)} + 3.746}{30.292}$$

The total estimated dry weight intake for each occasion was then
calculated by multiplying the estimated dry weight for each length
estimate by the number of worms of that length. Finally, an estimated
rate of dry weight intake was caculated by dividing by the length
of the observation period (min.) on that day.

The substrate temperature was measured at a depth of 5 cm. on
Seal Sands. Midday air temperatures were obtained from the weather
records of the Teesmouth coastguard station.

RESULTS

Counts

Curlews are present on the Tees estuary in large numbers from
July to April. Although very variable, the number of Curlews on Seal
Sands at low water fell to a minimum in mid winter between autumn
and spring peaks (Fig. 1). Similar mid-winter minima in estuary
counts have been found elsewhere in Britain (Wilson, 1973; Prater,
1973, 1974, 1976, 1977; Smith and Greenhalgh, 1977; Bainbridge and
Minton, 1978). However, counts made at high water show that the
total number of Curlews present on and around the Tees estuary
remained remarkably constant throughout the 1977/78 winter and from
September to December and February to April in 1976/77 (Fig. 1).
Some Curlews must, therefore, have moved from Seal Sands to alternative feeding areas in mid winter. This movement was not to Bran

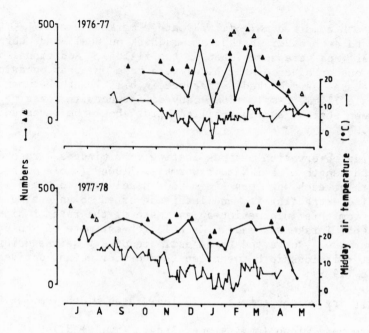

Figure 1. The relationship between counts of Curlews on Seal Sands
at low water (upper line) and the midday air temperature (lower line).
Also shown are counts at high water made by the author (solid
triangles) and the Birds of Estuaries Enquiry (open triangles).

Sands, the other large intertidal area on the estuary, as numbers
feeding there at low water remained below 25 until March (D.M.
Brearey, pers. comm.) and the changes in numbers there paralleled
those on Seal Sands.

    Instead, the changes in numbers of Curlews using Seal Sands,
not only in the long term but also from day to day, are caused by
the variable use of their other major feeding habitat, the low-lying
pastures adjacent to Seal Sands. Groups of up to 40 Curlews fed on
the Brinefields pasture, where most observations were made, and
movements between the mudflats and other fields were observed on
many occasions. The division of feeding Curlews between the two
habitats was determined primarily by the weather. Firstly, the
gradual decrease in autumn and increase in spring in the Seal Sands
low water counts paralleled the seasonal changes in air temperature
(Fig. 1). Secondly, in mid winter when the pastures were frozen
and/or covered with deep snow, e.g. 29 December 1976, 11 February
1978, Curlews were unable to feed on the fields and had to move back
to Seal Sands to feed; hence the high counts on Seal Sands on these
days. When the fields were waterlogged, due to a rapid thaw of snow,
e.g. 16 January 1977, heavy rain, e.g. 2 February 1978, or extensive

sea flooding, e.g. 11 January 1978, many earthworms came to the surface. Consequently, many Curlews moved to the fields, resulting in the very low counts on Seal Sands.

Although it is clear from these counts that the Curlew population on the Tees estuary used two habitats for feeding, observations on marked birds revealed that individual Curlews followed different patterns in their use of the fields and mudflats.

Marked Birds

Twenty-eight individually-marked Curlews were seen feeding on the pastures adjacent to Seal Sands during the three winters 1976-1979. All but one of these were also observed feeding on Seal Sands at some time during the study. Field-feeding must be a regular behaviour for at least some Curlews as 13 of these 28 birds (46%) were seen on the fields in more than one winter. Because of the differences in bias in data collection in the three winters, data for each winter are initially considered separately. Any individual seen in more than one winter is counted as a different bird in each.

The marked Curlews feeding on Seal Sands in early autumn followed one of at least three, and possibly four, basic patterns:

1. Some fed on the mudflats in autumn; then, as the temperature fell, they moved to feed exclusively on the adjacent fields in mid winter; some returned to feed on Seal Sands in spring. The low water sightings of the six marked Curlews known to follow this pattern of use of the two habitats in 1977/78 are shown in Fig. 2I. Six marked birds similarly moved to the fields as the temperature fell in the 1978/79 winter. These birds returned to feed on Seal Sands in mid winter only when the fields were frozen and covered in deep snow (see Fig. 2I).

During 1977/78 one marked Curlew which left Seal Sands after feeding there in autumn was later seen feeding on the fields, but only at high water. Its absence from Seal Sands (Fig. 2I) indicates that it too fed somewhere on the fields at low water during mid winter.

2. Other Curlews fed on Seal Sands at low water and, occasionally, used the fields to provide supplementary feeding over high water. Five marked birds showed this behaviour in 1976/77 and two in 1977/78 (Fig. 2II). Each bird was seen on Seal Sands at low water on the same day as it fed on the fields at high water. Three birds clearly fed on Seal Sands throughout the winter and used the supplementary feeding on the fields only occasionally (Fig. 2II). The other four were marked just before the March 1977 observations were made so their behaviour earlier in the winter is not known. As supplementary feeding on the fields was shown by several birds in

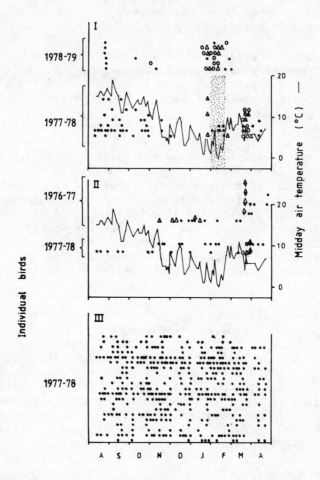

Figure 2. Sightings of marked Curlews on Seal Sands at low water
(filled circles), and on the fields at low water (open circles) and
high water (open triangles). Within each graph, each horizontal
line of symbols comprises all the observations of one individual
during one winter. Midday air temperature during the 1977/78 winter
is shown in the upper two figures.

I)   Birds which moved from the mudflats in autumn to feed
        exclusively on the fields until spring, except during deep
        snow or frozen ground (stippled).
II)   Birds which fed on Seal Sands at low water and used the
        fields only occasionally, for supplementary feeding at high
        water.
III)   Birds which fed exclusively on the mudflats throughout the
        winter.

March, in both winters, it may have been necessary for some birds
when the absolute density of <u>Nereis</u> on Seal Sands had fallen to a low
level. Alternatively, the extra feeding may have been required to
lay down fat reserves in preparation for migration. Supplementary
feeding did not occur only when the temperature was low and, therefore, energy requirements higher and <u>Nereis</u> less accessible.

3.   Some Curlews fed on Seal Sands throughout the winter and were
      never seen feeding on the fields at high or low water. The
number of Curlews in this category during 1977/78 greatly exceeds
those discussed above (Fig. 2I) because the majority of observations
were made of birds on Seal Sands as opposed to the fields.

4.   Less frequent sightings of other marked Curlews indicate that
      another pattern in use of the fields may be employed, namely
that some Curlews are purely field feeders, from autumn to spring.
During 1976/77 and 1977/78, three marked birds were seen feeding
on the fields at high water from November to March, but never on
Seal Sands during the same winter, despite the high frequency of
checks made on the mudflats.

      Finally, a few marked Curlews have shown the day-to-day varia-
tion in low water feeding site revealed in the low water counts for
Seal Sands (Fig. 1). Three birds fed on the fields instead of their
usual area, Seal Sands, on two days with strong winds (Townshend,
1981), which would have hindered the capture and handling of prey
on Seal Sands. (Feeding on the relatively sheltered fields would
also have reduced heat loss).

      The pattern of use of the fields followed by each individual
was related to its bill length, shorter-billed birds in general
spending more time on the fields. Figure 3i shows that the proportion of females on the fields was significantly lower than that on
the mudflats (fields: 6 females in 44; mudflats: 10 females in 22;
χ^2 = 8.09, 1 d.f., P < 0.01). Amongst the field feeders none of
those which fed there at low water had bills greater than 130mm,
and longer-billed birds were seen there only at high water (Fig.
3ii). A higher number of birds with very short bills (<115mm) fed
exclusively on the fields in mid winter (pattern 1) (8 out of 13)
than used the fields for supplementary feeding at high water (2 out
of 7) (Fig. 3iii), but the difference is not significant due to
the small number involved. The three marked birds believed to have
fed on the fields throughout their stay had extremely short bills
(mean = 109.3mm, S.E. 3.53). This was smaller, although not significantly so, than that of the 13 Curlews which moved to the fields
only in mid winter (mean = 114.1mm, S.E. 2.10).

      Thus, there is a correlation between the length of the bill and
the time spent on the fields during a winter. However, the range of
bill lengths of the birds feeding on Seal Sands throughout the winter

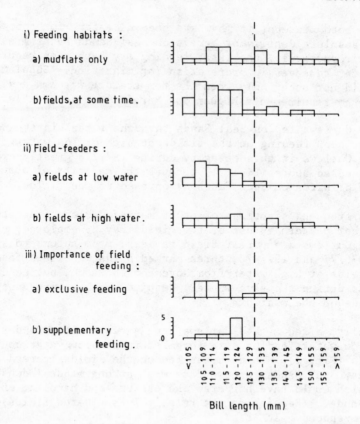

i) Feeding habitats :
   a) mudflats only
   b) fields, at some time.

ii) Field-feeders :
   a) fields at low water
   b) fields at high water.

iii) Importance of field
      feeding :
   a) exclusive feeding
   b) supplementary
      feeding.

Bill length (mm)

Figure 3. The distribution of bill lengths of marked Curlews seen
feeding on the fields and/or mudflats. The vertical axes indicate
numbers of birds.

i)   a) All those that fed exclusively on the mudflats through-
         out the winter.
     b) All those that fed on the fields at some time.
ii) a) Those from ii) that fed on the fields at low water.
     b) Those from ii) that fed on the fields only at high water.
iii) a) All those which fed exclusively on the fields in mid
         winter.
     b) All those which used the fields only for supplementary
         feeding.

(98-160mm) shows that, although some males moved to the fields, many
remained on Seal Sands under all conditions.

Foraging Behaviour

     The preponderance of males amongst the Curlews moving to the
fields can be related to differences between the sexes in their

foraging behaviour. The marked male Curlew, whose foraging on
Seal Sands was studied in detail, captured Nereis at as fast a
rate as the female at most temperatures (Fig. 4). However, at low
substrate temperatures the female, but not the male, maintained its
capture rate. As birds require more food at lower temperatures it
is, therefore, likely that the male would be the first to have
difficulty meeting its energy requirements during cold weather.

The estimated biomass intake rate of the female was considerably
higher than that of the male on every occasion (Fig. 4). As female
Curlews are larger than males they require more food per day. The
ratio of standard metabolic rates for males and females, calculated
from the equation for non-passerines in Lasiewski and Dawson (1967),
shows that males require about 12% less food per day than females
(see Townshend,1981). However, the estimated biomass intake rate
of the male Curlew on Seal Sands was only 33-55% that of the female,
falling to 10% in cold weather in January. The male Curlew probably
fed for longer in each tidal cycle (see Townshend,1981) and might
have fed more at night, but it is unlikely that this would have
increased the total biomass intake sufficiently.

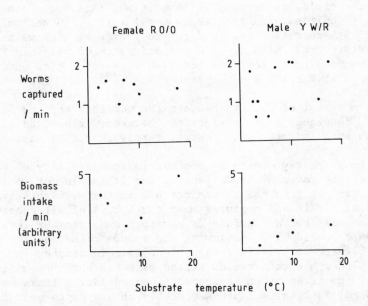

Figure 4. A comparison between the rates of capture of Nereis
worms and the estimated rates of biomass intake of a male and a
female marked Curlew which fed on the same mudbank on Seal Sands
during the 1977/78 winter. The biomass intake rates were measured
for both birds on the same 6 days.

DISCUSSION

Most of the Curlews present on the Tees estuary in autumn fed
on Seal Sands (Fig. 1), so those that later moved to the fields must
have done so because the relative profitability of feeding on these
two sites changed for these individuals. The fields become more
profitable when waterlogged but less so when frozen or covered with
deep snow; the mudflats are less profitable during gales. These
changes can occur within a period of days, and a few Curlews responded to them with similar rapidity. These short-term changes, however,
cannot account for the gradual decrease in the number of Curlews
feeding on Seal Sands in autumn. It appears that feeding on Seal
Sands, at least for short-billed (male) Curlews, becomes progressively more difficult as the temperature falls. It is known that
Nereis diversicolor lie deeper in their burrows at lower temperatures
(Muus 1967), and they must, therefore, move beyond the probing
reach of short-billed Curlews first. This accounts for the
difficulty experienced by the male Curlew Y W/R in attempting to
maintain its prey capture rate and biomass intake rate at low temperatures even though feeding alongside another bird that could do
so. Thus, it appears that at least some males, but not females,
have difficulty meeting their energy requirements on Seal Sands
during the coldest conditions of the winter. Some males, therefore,
move to the fields in autumn, perhaps in anticipation of the mid
winter difficulties, and feed there until spring. Other males,
together with some females, feed on the fields at high water when
supplementary food is needed. A preponderance of males in flocks
of Curlews feeding on pastures has also been noted elsewhere
(Elphick, 1979; Ens, 1979).

The movement of Curlews between the mudflats and the fields
was not a consequence of interactions between Curlews and other
waders as such interactions occurred rarely in either habitat.

Although partly due to a decrease in accessibility of Nereis
in winter, the movement of Curlews to the fields in autumn is
probably also in response to an increase in the availability of
earthworms. On inland pastures most earthworms of all species are
active and near to the surface in November and December (Evans and
Guild, 1947; Gerard, 1967) following the autumn rains, so their
availability to Curlews increases. During cold weather in mid
winter, although earthworms on inland pastures burrow deep in the
substrate and often become quiescent (Gerard, 1967), those on the
pastures adjacent to Seal Sands probably change depth very little.
The earthworms were found only in the thin surface layer of soil
and not in the underlying clay, and always on areas that generally
remained above the water table (Davidson, 1980). As a consequence
most would have remained within reach of the Curlew's bill (N.C.
Davidson, pers. comm.). Therefore, except when the ground was
frozen or covered in deep snow, the availability of earthworms was

adequate for male Curlews to feed on the fields even though most
Nereis were beyond their reach on the mudflats.

Some of the Curlews that moved to the fields when the temperature fell in autumn did not return to Seal Sands when the temperature increased in spring. Presumably the absolute density of Nereis
on Seal Sands had fallen, due to predation and natural mortality,
to such an extent that the spring temperature increase brought a
potential increase in foraging profitability which was insufficient
to make feeding there preferable to feeding on the fields. It is
probable that the increasing temperature in spring also increased
the availability of food on the fields by increasing the activity
of earthworms (Evans and Guild, 1947) and other invertebrates.

Detailed observations of marked Curlews on Seal Sands during
the winter have revealed that a minority of Curlews defend feeding
sites for periods of weeks or months (Townshend, 1981). Such long-
term territories are held only on certain areas of Seal Sands. None
of the marked Curlews seen feeding on the fields (at high or low
water) defended long-term territories. This may be because the
birds which feed on the fields are subordinate individuals and thus
unable to acquire and retain exclusive feeding sites on Seal Sands.
However, those birds which move to feed exclusively on the fields
in mid winter, may not attempt to acquire territories in autumn,
even though capable of doing so, because they will later be moving
to the fields; i.e. the costs of defence outweigh the benefits,
in the time available (Townshend, 1981). The fidelity of certain
individuals to field feeding in successive winters suggests that
the latter hypothesis may apply to at least some individuals.

Although males and females are distributed differently between
the two habitats, it seems unlikely that the sexual dimorphism in
Curlews has evolved to reduce intraspecific competition for food,
as has been suggested for raptors (Reynolds, 1972; Newton, 1979),
woodpeckers (Selander, 1966, Hogstad, 1978) and finches (Newton, 1967),
during the non-breeding season. The evidence presented above
indicates that the males moving to the fields do so because they are
having difficulty capturing Nereis fast enough on the mudflats in
cold weather, i.e. they are forced to move. Other males feed on the
mudflats all winter. If the dimorphism in bill length gave males
an advantage on fields (with the shorter, straighter male bill perhaps being better for probing into firm substrata and less likely
to break when twisted) and females the advantage on mudflats (with
the longer bill being more adept at capturing worms deep in soft
sediment) all males should feed on the fields throughout their stay
on the estuary, rather than just some doing so and only when
conditions deteriorate on the mudflats.

ACKNOWLEDGEMENTS

I am grateful to Professor D. Barker for the use of the
facilities of the Department of Zoology, University of Durham; to
Dr. P.R. Evans for providing supervision, criticism and encouragement; and to my colleagues N.C. Davidson, P.J. Dugan, L.R. Goodyer
and M.W. Pienkowski for helpful discussions during this research.
The sightings of marked birds during the 1978/79 winter were supplied
almost entirely by N.C. Davidson, L.R. Goodyer and F.L. Symonds.
The research was financed by a NERC Research Studentship.

REFERENCES

Bainbridge, I.P. and Minton, C.D.T. 1978. The migration and mortal-
    ity of the Curlew in Britain and Ireland. Bird Study 25: 39-50.
Bannerman, D.A. 1961. The Birds of the British Isles, Vol. IX.
    Oliver & Boyd, Edinburgh and London.
Davidson, N.C. 1980. Seal Sands Feasibility Study: An investigation
    into the possibility of developing alternative areas in com-
    pensation for the eventual loss of the remaining areas of Seal
    Sands, Teesmouth. Unpublished report to Cleveland County
    Council and the Nature Conservancy Council.
Elphick, D. 1979. An inland flock of Curlews Numenius arquata in
    mid-Cheshire, England. Wader Study Group Bulletin 26: 31-35.
Ens, B. 1979. Territoriality in Curlews Numenius arquata. Wader
    Study Group Bulletin 26: 28-29.
Evans, A.C. and Guild, W.J. 1947. Studies on the relationships bet-
    ween earthworms and soil fertility. I. Biological studies in
    the field. Annals of Applied Biology 34: 307-330.
Gerard, B.M. 1967. Factors affecting earthworms in pastures.
    Journal of Animal Ecology 36: 235-252.
Hogstad, O. 1978. Sexual dimorphism in relation to winter foraging
    and territorial behaviour of the Three-toed Woodpecker Picoides
    tridactylus and three Dendrocopus species. Ibis 120: 198-203.
Knights, P.J. 1979. Effects of changes in land use on some animal
    populations. Unpublished Ph.D. Thesis, University of Durham.
Lasiewski, R.C. and Dawson, W.R. 1967. A re-examination of the
    relation between standard metabolic rate and body weight in
    birds. Condor 69: 13-23.
Muus, B.J. 1967. The fauna of Danish estuaries and lagoons,
    distribution and ecology of dominating species in the shallow
    reaches of the mesohaline zone. Meddelelser fra Danmarks
    Fiskeri og Havundersøgelser 5: 1-316.
Newton, I. 1967. The adaptive radiation and feeding ecology of
    some British finches. Ibis 109: 33-98.
Newton, I. 1979. Population Ecology of Raptors. T. & A.D. Poyser,
    Berkhamsted.
Pienkowski, M.W. 1973. Feeding activities of wading birds and
    Shelducks at Teesmouth and some possible effects of further
    loss of habitat. Unpublished report to Coastal Ecology Research

Station, Institute of Terrestrial Ecology.

Prater, A.J. 1973. BTO/RSPB Birds of Estuaries Enquiry. Report for
    1971-72. Tring: British Trust for Ornithology.

Prater, A.J. 1974. Ditto, Report for 1972-73.

Prater, A.J. 1976. Ditto, Report for 1973-74.

Prater, A.J. 1977. Ditto, Report for 1974-75.

Prater, A.J., Marchant, J.H., Vuorinen, J. 1977. Guide to the
    identification and ageing of Holarctic Waders. BTO Guide 17.

Ratcliffe, P.J. 1979. An ecological study of the intertidal
    invertebrates of the Humber estuary. Unpublished Ph.D. Thesis,
    University of Hull.

Reynolds, R.T. 1972. Sexual dimorphism in accipiter hawks: a new
    hypothesis. Condor 74: 191-197.

Selander, R.K. 1966. Sexual dimorphism and differential niche
    utilization in birds. Condor 68: 113-151.

Smith, P.H. and Greenhalgh, M.E. 1977. A four-year census of wading
    birds on the Ribble estuary, Lancashire/Merseyside. Bird Study
    24: 243-258.

Townshend, D.J. 1981. The use of intertidal habitats by shorebird
    populations, with special reference to Grey Plover (Pluvialis
    squatarola) and Curlew (Numenius arquata). Unpublished Ph.D.
    Thesis, University of Durham.

Wilson, J. 1973. Wader populations of Morecambe Bay, Lancashire.
    Bird Study 20: 9-23.

Witherby, H.F., Jourdain, F.C.R., Ticehurst, N.F. and Tucker, B.W.
    1945. The Handbook of British Birds, Vol. IV. H.F. & G.
    Witherby, London.

MIGRATION AND DISPERSAL OF SHOREBIRDS AS A SURVIVAL STRATEGY

P.R. Evans

Department of Zoology
University of Durham
South Road, Durham

INTRODUCTION

The life-history strategy adopted by individuals of all wader
(Charadrii) species,to maximise their lifetime reproductive outputs,
is that of repeated reproductive attempts, each involving relatively
small parental investments. In the non-breeding season, they
usually migrate away from their breeding areas, sometimes over
distances of more than 10,000 km. Different individuals of the same
species may migrate to different destinations.

Ten species of waders occur regularly on British estuaries in
winter (Table 1). Another two, the Lapwing Vanellus vanellus
and the Golden Plover Pluvialis apricaria also spend the winter
on estuaries, in varying numbers according to the weather, but
much larger numbers winter inland. Two of the ten, the Curlew
Numenius arquata and the Redshank Tringa totanus, exploit coastal
pastures as well as estuaries in winter, the Curlew sometimes
moving much further inland (Elphick, 1979). These four species
breed chiefly well inland across northern temperate regions,
though Redshank also nest on salt-marshes. Of the other eight
species,only two breed in large numbers on the sea-coast itself,
the Oystercatcher Haematopus ostralegus and the Ringed Plover
Charadrius hiaticula, and then only in northern temperate regions,
where the Oystercatcher also breeds inland. Five species
confine their breeding chiefly to land north of the arctic circle
(Grey Plover Pluvialis squatarola, Turnstone Arenaria interpres,
Bar-tailed Godwit Limosa lapponica, Knot Calidris canutus and
Sanderling Calidris alba), whilst the arctic breeding populations
of Ringed Plover and Dunlin Calidris alpina outnumber those breeding
in the northern temperate zone (Pienkowski, 1980a; Pienkowski and

275

Table 1.   The More Abundant Wading Birds (Charadrii) on British
           Estuaries

(Average January counts 1974-78)

Species	Numbers (thousands)	Main Breeding areas
Dunlin	560	W. Siberia
Oystercatcher	200	Norway/Iceland/Britain
Knot	190	Greenland/Canadian arctic
Redshank	80	Iceland/Britain
Curlew	60	Fennoscandinavia/Britain
Bar-tailed Godwit	40	W. Siberia
Grey Plover	14	W. Siberia
Turnstone	10	Greenland/Canadian arctic
Sanderling	10	W. Siberia (Greenland?)
Ringed Plover	8	Britain/Scandinavia

Source: A.J. Prater (pers. comm.)

Dick, 1975). It will be clear, therefore, that most individuals of
most species change their feeding habitats between the breeding and
non-breeding seasons, though the distances they move to accomplish
this vary between species, and also amongst the various geographical
populations within a species. They cope with the different prey
spectrum available in different areas by specialising in foraging
method, rather than in prey species (Pienkowski, this volume).

The importance of migration (here defined as a directed seasonal
movement of a population, as opposed to a dispersal, which implies
'spreading-out' without a preferred direction) in the life-cycles
of arctic breeding species is obvious, since their inland breeding
grounds are covered by snow and ice during the winter and the
invertebrate foods upon which they live become inaccessible. It is
not self-evident why so many wader species breed, and at high densities, in arctic regions, particularly since their chicks (which feed
themselves from shortly after hatching, except in the case of the
Oystercatcher) may grow no faster there than in temperate regions,
as has been shown for the Ringed Plover by Pienkowski (1980a). It
has been claimed that greater habitat diversity at high latitudes
leads to a greater variety of breeding waders there (Järvinen and
Vaisänen, 1978). This may be true within Finland, but if applied

to larger geographical units, e.g. central or southern Europe, it
is possible to find an even greater variety of habitats within a
defined area without comparable increases in the number of breeding
wader species. Järvinen and Vaisänen dismissed the role of predation
as a factor affecting the variety of species since they could find
no evidence of higher predation rates at higher latitudes. Apparently
they were arguing by analogy with rocky-shore intertidal communities,
in which increased predation pressure allows the co-existence of a
greater variety of prey by preventing competitive exclusion (Paine,
1974). There is little evidence that competition between breeding
wader species would occur in the absence of predation. What is more
important, however, is that predation _decreases_ at higher latitudes
(Larson, 1960; Pienkowski, 1980a), so that hatching and fledging
success of wader chicks are greater in the arctic, e.g. greater in
Greenland than in Britain. Pienkowski suggests that the southern
limit of distribution of Ringed Plovers in Europe (and, by implication, also of other species) is set by increasing predation pressure
at lower latitudes.

The major disadvantage of migrating to the arctic in spring is
the lack of predictability of the date of snow melt (Green, Greenwood
& Lloyd, 1977). On this depends the exposure of the sites upon which
the birds breed and feed. Because such areas become unsuitable for
birds only a few weeks later, in late summer, delays in the date of
snow melt may make it hazardous and unproductive for birds to attempt
to breed in some years. The life-history strategies of arctic-breeding
waders, therefore, require their potential for reproduction to be
spread over many years. This in turn demands the selection of
migration strategies and of non-breeding quarters in which the chances
of survival are high.

In contrast to the arctic species, it is both possible and
profitable for waders breeding in the northern temperate regions to
attempt to nest each year. However, losses of whole clutches of
eggs and of chicks, to both avian and mammalian predators, occur
with such intensity in some years that annual production of fledged
young from a given locality is, on average, low (Pienkowski, 1980a).
Thus, for individual adults of these populations also, it is
important to select "wintering" areas in which the chances of survival are high.

In this paper, I shall show first that annual survival of adults
of even the smallest of the arctic breeding species is usually high;
I shall then consider the meteorological conditions in the wintering
areas that make this possible and attempt to define geographical
limits within which species could winter. This approach leads to a
discussion of the balance between longer migrations and less vulnerable wintering areas, and finally to the possibility that within a
single species two strategies may be found in the use of wintering
areas, namely, residence and transience.

SOURCES OF INFORMATION

Since the autumn of 1975, about 10,000 waders have been caught and banded with numbered metal rings on the Tees estuary in north-east England (54°37'N, 1°12'W). Additionally, a few hundred of each of Grey Plover, Turnstone, Curlew and Sanderling have been marked with unique combinations of colour-rings, so that their subsequent survival could be followed by direct observation. Knot, in contrast, have been colour-dyed with a variety of colours on their underparts, to enable movements within Britain during a winter to be monitored. (These colours are lost when the feathers are moulted in the spring). Further information on movements has been gathered from the annual reports of the British Trust for Ornithology's bird-ringing scheme and from details of the results of this scheme published in success-ive numbers of the Wader Study Group's Bulletin. Other sources of information are acknowledged in the text.

ANNUAL SURVIVAL RATES

Sightings in successive autumns or winters of colour-marked birds on the Tees estuary have provided minimum annual survival rates for five species (Table 2). Since some individuals may stay for only a short period on the estuary each year, and so escape observations during the weekly censuses (sometimes hindered by bad weather), and since the possibility exists that a few individuals may change the estuaries they use between years, these figures must be regarded as minima. In each case they refer to adult birds, and few data are available from juveniles for comparison. The most extensive concern 13 juvenile Turnstone marked in winter 1976/77, of which 10 (77%) were seen the following winter. The cause of their survival rate being lower than that of adults, whether in-creased mortality occurred in winter, on migration, or during their first summer of life, is not known.

The values for 1975-78 emphasize that annual survival is usually high even in species undertaking long migrations to and from their breeding areas (Grey Plover, Bar-tailed Godwit and probably most Sanderling to western Siberia; Turnstone and perhaps some Sanderling to Greenland and eastern arctic Canada). This indicates that the increases in energy reserves before migration to and from Britain, and the choice of departure conditions, are well adapted to the length of the journeys and the routes followed. Survival on the breeding grounds themselves must also be high, with losses of incuba-ting birds (as opposed to their eggs and chicks) to ground predators being low. Occasionally, however, heavier mortality of adults does occur on the breeding areas when weather becomes severe just after the birds have arrived (e.g. Morrison, 1975; Marcstrom and Mascher, 1979). Nevertheless, in most years the greater part of the annual mortality occurs in the "wintering" areas, although, in mild winters, even this mortality is low, as illustrated by the 1975-78 survival

Table 2. Minimum Annual Survival Rates of Shorebirds Visiting the Tees Estuary during the Non-Breeding Season

Species	1975 to 1978 inclusive			1978 to 1979		
	No. marked in autumn y	No. seen in autumn y+1	% survival	No. marked in autumn 1978	No. seen in autumn 1979	% survival
Grey Plover	71	57	80%	29	19	66%
Turnstone	184	162	88%	66	55	83%
Curlew*	119	98	82%	61	46	75%
Bar-tailed Godwit	8	8**	100%	8**	7	88%
Sanderling	93	85	91%	55	37	67%

* a few Curlew are shot by wildfowlers at Teesmouth each winter; survival rates might otherwise be higher.

**includes one bird seen only on the Humber estuary in autumn 1978 but which had returned to Teesmouth in autumn 1979.

figures at Teesmouth, one of the coldest estuaries in Britain
(see Fig. 2 in Pienkowski et al., 1979).

The considerable increase in mortality rates in the year between
autumn 1978 and autumn 1979 may be attributed chiefly to the severe
weather conditions during the first three months of 1979, when corpses
of a few marked Sanderling and Grey Plover were found on the tide-
line at Teesmouth (a rare occurrence in other winters). Even in
this winter, evidence for birds dying as a result of low temperatures,
acting through reduced prey availability and increased daily food
requirements, was slight. It has been shown by Dugan et al., (in
press) for the Grey Plover, and for certain other species by
Davidson (this volume), that the occurrence of gales, preventing birds
from feeding for several successive days during the cold weather,
led to much greater depletion of their energy reserves than did low
temperatures alone. Indeed, spells of several days of cold weather
without gales occurred in the previous winters without leading to
heavy mortality, and the accounts of the effects of the severe
winters of the 1880's and late 1910's by Chapman (1907, 1924) leave
no doubt that most shorebirds at Lindisfarne, Northumberland,
survived weather colder than that of recent years without loss of
condition ("even in our most severe seasons, the knots remain fat
as butter" (Chapman, 1924)).

FACTORS AFFECTING THE DISTRIBUTION OF SHOREBIRDS IN WINTER

In order to survive until the next breeding season, birds must
move to non-breeding areas where they can feed sufficiently fast,
for sufficiently long each day, to meet their average daily energy
and protein requirements. (The role of fat and protein reserves in
aiding survival over short periods - several days - of poor feeding
conditions has been discussed elsewhere, (Evans & Smith, 1975;
Davidson, this volume)).

The rate at which birds can feed depends in part on the density
of available prey, which in turn depends upon the absolute density
of prey and the proportion that is available at any moment. Availability can be assessed only in relation to the foraging technique
employed by the particular shorebird predator. Only those infauna
of soft sediments that lie at depths within range of a bird's probing
bill are available to those predators that feed by touch, e.g.
Curlew (Townshend this volume). Only those that are present and
active close to the surface of soft sediments are available to birds
feeding by sight e.g. plovers (Pienkowski this volume), and only
those sessile invertebrates, usually of rocky shores, that expose
their soft parts, e.g. through gaping shells, or fail to adhere
sufficiently powerfully to the substratum, are available to birds
that wish to attack them.

The meteorological factors that most severely affect prey availability are temperature, wind and rain, the last-named by reducing the detectability of movements of prey at the surface of soft substrates. In general, the activity of intertidal invertebrates (which affects how often they come to the surface and are detectable) is depressed by decreasing temperatures, particularly below about $6^{\circ}C$, and by increasing wind speeds which lead to more rapid drying of the surface (Pienkowski, 1980a; Smith, 1975). The infauna also tends to be found at greater depths in cold weather.

Decreases in the percentage of prey available at any one moment can be counteracted by birds concentrating into areas of higher absolute density of prey, in order to maintain their rate of food intake (Pienkowski, 1980a). However, prey availability may also be decreased by the movements of the predators themselves and hence is related to the density of birds in the feeding area (Goss-Custard, 1970). Additionally, as the density of predators increases, they may come to interfere with each other's attempts to capture prey, so that even if the density of available prey were to remain constant, the rate at which each predator could capture a prey item would decrease (Pienkowski, 1980a). In spite of all these effects, in most British winters, mortality resulting from the normal seasonal fall in temperature to a January monthly average of no less than $3^{\circ}C$ on the coast (in eastern Scotland) is slight, since birds employ a variety of methods to ameliorate the effects of short cold spells (Davidson, this volume). Emigration in direct response to such spells is rare in shore waders (Evans, 1976).

Insofar as winter temperature sets the outer limits to the northern and eastern parts of the European winter range of those wading birds that feed by probing, it probably does so by the regularity with which ice forms over, and binds to, the surface layers of muddy sediments during low water periods. When the tide floods again, the surface layers and the invertebrates they contain may then be shifted to other areas and many killed. The low densities of Corophium volutator in the upper layers of the inter-tidal sediments of the Bay of Fundy, North America, in spring have been attributed to severe ice scour in winter (Yeo, 1977 quoted in Hicklin and Smith, 1979); consequently many fewer birds use the area as a staging post on migration in spring than in autumn, when Corophium densities reach several tens of thousands per m^2. No shorebirds winter there. A parallel example in Europe could be the failure of many waders, notably Dunlin and Knot, to use the Baltic shores as refuelling sites on spring migration. Both these species feed from the upper layers of the mud, on small items, including annelids, crustacea and molluscs, that probably do not survive transportation in frozen sediment during ice-scour. Because of its salinity, the shores of the Baltic freeze more easily than those around the North Sea, and although they support many tens of thousands of Dunlin and Knot in autumn, in spring these two species

use the Schleswig-Holstein section of the Wadden Sea as a major
refuelling site on the migration route to western Siberia (Dick,
1979).

Ice-scour probably occurs rather infrequently on the North Sea
coasts and estuaries, but temporary freezing of the surface layer
of sediment is more regular in mid-winter. Since mud surfaces,
which retain water, are more likely to freeze than sandy sediments,
it would be expected that the smaller waders that feed by probing
in muddy sediments would concentrate in slightly milder areas in
winter than the larger species that feed either on sand flats, or
in the water at the tide edge. Table 3 provides some support for
this contention, since higher proportions of the European wintering
population of Curlew, Bar-tailed Godwit and Oystercatcher than of
Dunlin and Redshank are found in the Wadden Sea (chiefly in Germany
and the Netherlands) than around the milder coasts of Britain.
(Although Knot and Oystercatchers feed on Macoma balthica and
Cerastoderma edule from the upper layers of the Wadden Sea sediments,
these are primarily sandy; and many Oystercatchers feed on mussels
Mytilus edulis, which are attached to the surface of the substratum).

The plovers which feed visually, may be more directly limited
in their winter distribution by low temperatures just above freezing
point, acting through their depressive effect on prey activity, as
has been argued convincingly by Pienkowski (1980a, 1980b). Within
their normal wintering area, during the winter of 1978/79, it was
only the coincidence of high winds with low temperatures that caused
Grey Plovers to lose condition (Dugan et al., in press) and in some
cases to die. Since strong winds can prevent feeding by all the
long-legged waders, Redshank, Grey Plovers, Curlew and Bar-tailed
Godwit (Davidson, this volume), and since wind strength tends to
vary seasonally, being highest between November and January in north-
east England (Dugan et al., in press), it is possible that any
geographical pattern of variation in wind strength could combine
with the effects of temperature to determine the limits of winter
distribution of these four wader species.

It was thought formerly (Evans, 1976) that daylength was
important in restricting the time for which some shorebirds could
feed effectively during each 24 hours. However, it has now been
shown that even for many visual foragers, e.g. plovers, feeding at
night is both possible and profitable (Dugan, this volume). Hence
the time of high water in relation to dawn and dusk may not alter
the potential duration of foraging as much as was formerly suspected.
More critical, however, may be the weekly alternations between spring
and neap tides, and the tidal range itself, which alter the variety
of potential prey (greater on spring tides), and the total times
of exposure of those parts of the intertidal area in which foraging
is profitable.

Table 3. Proportions of Western European Wintering Populations of Certain Waders in Britain and the Wadden Sea

Species	Britain	Wadden Sea
Chiefly tactile foragers in upper 4 cm of sediment:		
Dunlin	43% (mud)	16% (mud)
Knot	65% (mud)	16% (mud)
Redshank	69% (mud)	15% (mud)
Chiefly tactile foragers in lower layers of sediment:		
Curlew	34%	45%
Bar-tailed Godwit	44%	27%
Visual and tactile forager in upper layer of sediment:		
Oystercatcher	30%	47%

Source: Average January counts 1974-78 (A.J. Prater, pers. comm.)

In estuaries with a small tidal range, strong onshore winds can severely reduce the duration of exposure in each tidal cycle of normally intertidal areas. For example, in the Wattenmeer in August 1980, westerly gales assisted the incoming tide to reach the predicted high water mark three hours before the actual time of high water, and Dunlin were forced to feed on the salt marsh (M.W. Pienkowski, pers. comm.). This effect will be less serious for the larger longer-legged waders, most of which do not need to feed for so long in each tidal cycle (Evans, 1980) and can feed in deeper water than the smaller species. In western Europe, the tidal range increases southwards from western Norway through north Germany, the Netherlands and Belgium to reach its greatest ampli-tude in northwest France, before decreasing through western France and northern Spain to a low amplitude in southern Portugal. In almost all parts of Britain, and particularly on the west coast, the tidal range is considerably greater than in the Wadden Sea, so that the duration of exposure of British intertidal flats is less influenced by wind strength. Another reason for the low proportion of Dunlin, Knot and Redshank wintering in the Wadden Sea (Table 3) may, therefore, be to avoid the shortened feeding periods associated with midwinter gales from the north-west, which hold in the tide.

THE TRADE-OFF BETWEEN THE LENGTH OF A MIGRATION AND THE CHANGES OF
SURVIVAL IN THE "WINTERING" AREA

   Although, on average, air temperatures are lowest in January,
sea temperatures lowest in February, and wind strengths greatest
in mid-winter, there is much year-to-year variation. It is rare
in Britain for severe cold and several successive days of gale-force
winds to coincide, as they did in the first three months of 1979.
Thus, in many winters, survival should be no problem for shorebirds
in areas to the north and east of the range in which they are normally
found. Conversely, their chances of survival as related to temperature would be even greater if they wintered further south in Europe.
It is not clear why more do not fly the extra few hundred kilometers
from their breeding grounds to achieve this. The energetic costs
might be too great, though this seems improbable. Possibly the
absolute densities of suitable prey on the Atlantic coasts of Spain
and Portugal will be found to be low, because of the coarseness of
sediments on beaches away from the few estuaries; and perhaps the
small tidal ranges there (similar to the Wadden Sea) mean that the
period of exposure of intertidal feeding areas may be seriously
shortened by the occurrence of onshore (westerly) winds. These
factors would act against successful overwintering by small sandpiper-
type waders, but not so markedly against plovers, which are able to
exploit low densities of prey if temperatures remain above about 10ºC
(Pienkowski, 1980b). Many waders do migrate further south, to the
Banc d'Arguin, Mauritania, but in 1973 some of these arrived in an
exhausted and emaciated condition (Dick and Pienkowski, 1979). Unlike
those arriving in western Europe from Siberia, the waders reaching
Mauritania must have completed the last part of their journey with
a long flight, since there are few suitable intertidal feeding areas
to the south of the Atlantic coast of France, except for certain
coastal sites in Morocco, where they could refuel. It would be
valuable to know how often, if at all, the long overwater flights
from Greenland to Europe lead to equal loss of condition to that
found in Mauritania. At least for adult Turnstone (Table 2), it
seems unlikely that any loss in condition resulting from migration
leads to mortality, since annual survival rates are normally high.

   I have suggested elsewhere (Evans, 1979a) that there may be an
advantage to individuals of certain species to winter in Europe as
close to the northern and eastern limits of their non-breeding range
as possible, so that they can then move back to pre-migration
"fattening grounds" which cannot be used safely by waders in winter,
and therefore where food has not been severely depleted during the
winter, as soon as these areas become habitable in spring. Such
individuals are, of course, more likely to die in the occasional
severe winters.

   It appears that this may occur in the Grey Plover, in which
some colour-marked individuals are resident on the Tees estuary,

close to the northern limit of their winter range, throughout the
non-breeding season, whereas other marked individuals return predict-
ably in January or early February of each year, having spent the
early winter elsewhere, probably further south (Townshend, 1981;
Dugan, 1981). We do not know at present whether these individuals
are in better condition when they return to Teesmouth than those
that have stayed throughout the early winter. Nor do we understand
why they return so early in the year, when feeding conditions would
appear to be poor, unless there may be changes in the nocturnal
activity patterns of their prey at that season which make the birds'
foraging more profitable than earlier in the winter (see Dugan, this
volume). This point is under investigation. Apparently both groups
of birds leave the Tees in March, and at least one colour-marked bird
has been seen in Schleswig-Holstein in April (en route for Siberia)
in an area where Grey Plovers do not overwinter. It is not yet clear
whether the winter "residents" on the Tees leave in spring, on
average, earlier than the "immigrants".

RESIDENCY OR TRANSIENCY - ALTERNATIVE STRATEGIES?

Most estuarine waders have to cope with diurnal/nocturnal and
tidal rhythms in the availability of their prey, and their behaviours
are adapted so that they feed in the more profitable areas (Evans,
1979b; Dugan, Pienkowski, this volume). Some species also show
adaptations to changes in the relative profitability of feeding on
the estuary or in hearby fields, e.g. Curlew (Townshend, this volume).
These changes in prey availability are reasonably predictable, and
the birds' responses are correspondingly refined. However, some
changes in the density of available invertebrates arise at unpredictable times, not from changes in the proportion of prey available, but
in their absolute density. These abrupt changes result from movement
of sediments by wave action. Unlike ice-scour, which is likely to
affect large stretches of coastline fairly uniformly, scour of sediments by wave action can be localized in nature and extent, depending
on the variable configuration of the shore in relation to a uniform
direction of wave movement dictated by the wind. Scour is likely
to be most severe on coastal sand beaches, where the profile can
alter markedly within a few days by a combination of wave action and
blown sand. Such alterations in profile may result in the removal
of extensive populations of prey species such as the polychaete
Nerine cirratulus, which lives in the upper layers of the sand.
This polychaete is an important food of Sanderling in north-east
England (D. Brearey, in prep.) and the unpredictability of its
continuing presence on sea beaches at Teesmouth may be the cause of
variations in the behaviour of Sanderling. Some Sanderling reach
Teesmouth in late July, moult their flight feathers, and remain there
throughout the winter until they migrate to the breeding areas in the
following May. Some marked individuals are seen in autumn and again
in spring, but not in winter; yet others are seen irregularly through
the winter, but not as often as expected. Some ringed birds have been

sighted feeding in winter on a beach some 30 km north of the estuary,
at Whitburn. Numbers counted on the beaches at Teesmouth fluctuate
from week to week in a way that is uncorrelated with food abundance
or availability, except that largest numbers are present when beds
of wrack (chiefly Laminaria) are washed up after onshore gales.
Similar opportunistic feeding behaviour of some individual Sanderling,
apparently as a result of the patrolling of several tens of kms of
coast, has been recorded in California (Myers and McCaffery, 1980).
It seems possible that there are at least two strategies for survival
in Sanderling, in response to the unpredictable nature of their main
food resources, namely residency (which requires that if their main
prey are swept away with a change in beach profile, they find alternative foods, or starve) and transiency (which requires a continual
assessment of food resources along many kilometers of beach, so that
if one area is scoured, another can be chosen for foraging).

Sediments may also be shifted from intertidal areas within
estuaries, both by wave action in the mouths of wide estuaries,
where sufficient fetch exists to allow high waves to build up, and
further up-river by flash floods. An annual cycle of deposition
and erosion on the Humber mudflats has been described by Ratcliffe
(1979), who also mentions much shorter-term changes in sediment
depth as a result of strong wave action. Sediments may be stabilized
in some sites by tube-building oligochaetes and polychaetes, e.g.
at Teesmouth (Kendall, 1979), but this stability may be broken down
if the worm populations crash (Fager, 1964). On the Humber, Tasker
(1979) described how, after gales, piles of dead bivalves (Macoma
balthica and Cerastoderma edule) built up along the tide line, where
they were scavenged by Turnstones. Removal of the surface-dwelling
invertebrates, by sediment shifts in an estuary, could have similar
effects to ice-scour, and might be expected to lead to changes in
foraging locations of bird species that feed from the upper layers
of the mud or sand.

Many adult Dunlin are known to be faithful to the same estuary
in successive winters. In large estuaries, such as The Wash, sub-
populations may be recognized in different parts, between which
rather little exchange of birds occurs (Minton, 1975). Individuals
do not normally change their feeding and roosting areas by more than
a few kilometers, as has been shown by colour-dyeing birds on the
Forth (Pienkowski and Clarke, 1980). At first sight, this situation
does not accord with the prediction that birds feeding from the
upper layers of the sediment should move in response to shifts of
sediment. However, at least on the Tees (Evans et al., 1979),
Dunlin feed chiefly on those abundant small annelid species that
stabilize the sediments. Hence only after very severe wave action
would their food supplies be removed.

In contrast, Knot, which feed chiefly on small bivalves and
Hydrobia ulvae, move about extensively between British estuaries in

winter (Prater, 1974; Dugan, 1981). The picture is complicated by
the choice of only a few large estuaries as moulting grounds by adult
birds returning in late summer. Many Knot renew their flight feathers
on The Wash (Lincolnshire/Norfolk), and disperse from there to other
estuaries in late October when moult is completed. This dispersal
is not caused by food shortage on The Wash, since some individuals
stay there throughout the winter and even more birds arrive (after
moulting elsewhere) than depart during October and November (Prater,
1979). It is not clear whether the dispersal from The Wash has
strong directional tendencies. Undoubtedly, many birds move north
up the east coast of England and into southeast Scotland (Dugan,
1979). However, others may cross to the west coast of Britain, in at
least some years. Once Knot have reached other estuaries e.g. the
Tees, only some individuals stay, others moving on within a few
days, both northwards to the Forth and even some southwards to the
Humber, as has been shown by sightings of birds dyed on the Tees.
Others may replace them on the Tees. As with the Sanderling, it
appears that individual Knot may be classed either as residents or
as transients after they have dispersed to the smaller British
estuaries from their moulting grounds. The movements undertaken by
the transients seem to be much more extensive than those of Sander-
lings, but may serve the same purpose, to identify a number of
potentially suitable feeding sites, in case the food resources of
one site are removed by wave action dislodging the small molluscs
which form their principal foods. In order that the 'transient'
strategy may be successful, birds should maintain a sufficiently
high level of fat reserves so that they can move to another site
whenever necessary, whereas 'residents' may not do so. The proportion of birds of each species using the alternative strategies should
relate to the chances of 'residents' starving as a result of catas-
trophic removal of their food resources.

ACKNOWLEDGEMENTS

    I am most grateful to M.W. Pienkowski, P.J. Dugan and N.C.
Davidson for comments on the first draft of this paper and for
information bearing on some of the speculations I have presented.
Catching and marking of shorebirds at Teesmouth has been directed
by L.R. Goodyer, who also made many of the subsequent observations
of colour-marked Sanderling and Turnstone. Other persons contribut-
ing to the research programme and results quoted in this paper
include D.M. Brearey and D.J. Townshend. I thank all six
most sincerely for their help. Many other students from Durham
University have assisted in the shorebird marking programme; their
contributions were invaluable. A.J. Prater kindly supplied information on winter counts of waders in Europe, in his capacity as co-
ordinator of wader studies for the International Waterfowl Research
Bureau.

    Financial assistance for parts of this study was provided by

the Nature Conservancy Council and the Nuffield Foundation, to whom
grateful thanks are due.

REFERENCES

Chapman, A. 1907. "Bird-life of the Borders"; 2nd Edition.
London : Gurney and Jackson.
Chapman, A. 1924. "The Borders and Beyond". London : Gurney and
Jackson.
Dick, W.J.A. 1979. Results of the WSG project on the spring migration
of Siberian Knot Calidris canutus 1979. Wader Study Group
Bulletin 27: 8-13.
Dick, W.J.A. and Pienkowski, M.W. 1979. Autumn and early winter
weights of waders in north-west Africa. Ornis Scandinavica
10: 117-123.
Dugan, P.J. 1979. Inter-estuarine movements of shorebirds. Wader
Study Group Bulletin 27: 19.
Dugan, P.J. 1981. Seasonal movements of shorebirds, in relation to
spacing behaviour and prey availability. Unpublished Ph.D.
thesis, University of Durham, U.K.
Dugan, P.J., Evans, P.R., Goodyer, L.R. and Davidson, N.C. (in
press). Winter fat reserves in shorebirds: disturbance of
regulated levels by severe weather conditions. Ibis.
Elphick, D. 1979. An inland flock of Curlews in mid-Cheshire,
England. Wader Study Group Bulletin 26: 31-35.
Evans, P.R. 1976. Energy balance and optimal foraging strategies in
shorebirds: some implications for their distributions and movements in the non-breeding season. Ardea 64: 117-139.
Evans, P.R. 1979a. Some questions and hypotheses concerning the timing of migration in shorebirds. Wader Study Group Bulletin
26: 30.
Evans, P.R. 1979b. Adaptations shown by foraging shorebirds to
cyclical variations in the activity and availability of their
intertidal invertebrate prey. pp 357-366 in "Cyclic phenomena in marine plants and animals (E. Naylor and R.G. Hartnoll,
Eds.) Oxford: Pergamon Press.
Evans, P.R. 1980. Reclamation of intertidal land: some effects on
Shelduck and wader populations in the Tees estuary. Verhandlung
orn. Ges. Bayern 23 (in press).
Evans, P.R., Herdson, D.M., Knights, P.J. and Pienkowski, M.W. 1979.
Short-term effects of reclamation of part of Seal Sands,
Teesmouth, on wintering waders and Shelduck. Oecologia 41:
183-206.
Evans, P.R., and Smith, P.C. 1975. Studies of shorebirds at
Lindisfarne, Northumberland. 2. Fat and pectoral muscle as
indicators of body condition in the Bar-tailed Godwit.
Wildfowl 25: 64-76.
Fager, E.W. 1964. Marine sediments: effects of a tube building
polychaete. Science (N.Y.) 143: 356-359.

Goss-Custard, J.D. 1970. Dispersion in some over-wintering wading
    birds. pp 3-35 in "Social behaviour in birds and mammals"
    (J.H. Crook (Ed.)) London: Academic Press.
Green, G.H., Greenwood, J.J.D. and Lloyd, C.S. 1977. The influence
    of snow conditions on the date of breeding of wading birds in
    north-east Greenland. J. Zool. (Lond.). 183: 311-328.
Hicklin, P.W. and Smith, P.C. 1979. The diets of five species of
    migrant shorebirds in the Bay of Fundy. Proc. Nova Scotia
    Inst. Sci. 29: 483-488.
Järvinen, O. and Vaisänen, R.A. 1978. Ecological zoogeography of
    North European waders, or why do so many waders breed in the
    North? Oikos 30: 495-507.
Kendall, M.A. 1979. The stability of the deposit feeding community
    of a mud-flat in the River Tees. Estuarine Coastal Marine Sci.
    8: 15-22.
Larson, S. 1960. On the influence of the Arctic Fox Alopex lagopus
    on the distribution of arctic birds. Oikos 11: 276-305.
Marcstrom, V. and Mascher, J.W. 1979. Weights and fat in Lapwings
    and Oystercatchers starved to death during a cold spell in
    spring. Ornis Scandinavica 10: 235-240.
Minton, C.D.T. 1975. Waders of The Wash - ringing and biometric
    studies. Report to NERC of Scientific Study G of the Wash
    Wader Storage Scheme Feasibility Study. 40pp.
Morrison, R.I.G. 1975. Migration and morphometrics of European
    Knot and Turnstone on Ellesmere Island, Canada. Bird-Banding
    46: 290-301.
Myers, J.P. and McCaffery, B.J. 1980. Opportunism and site-faithful-
    ness in wintering Sanderlings. (Abstract) Wader Study Group
    Bulletin 28: 43.
Paine, R.T. 1974. Intertidal community structure. Experimental
    studies on the relationship between a dominant competitor and
    its principle predator. Oecologia 15: 93-120.
Pienkowski, M.W. 1980a. Aspects of the ecology and behaviour of
    Ringed and Grey Plovers. Unpublished Ph.D. thesis, University
    of Durham, U.K.
Pienkowski, M.W. 1980b. Differences in habitat requirements and
    distribution patterns of plovers and sandpipers as investigated
    by studies of feeding behaviour. Verhandlung Orn. Ges. Bayern
    23 (in press).
Pienkowski, M.W. and Clark, H.A. 1979. Preliminary results of winter
    dye-marking on the Firth of Forth, Scotland. Wader Study Group
    Bulletin 27: 16-18.
Pienkowski, M.W. and Dick, W.J.A. 1975. The migration and wintering
    of Dunlin Calidris alpina in north-west Africa. Ornis
    Scandinavica 6: 151-167.
Pienkowski, M.W., Lloyd, C.S. and Minton, C.D.T. 1979. Seasonal
    and migrational weight changes in Dunlin. Bird Study 26:
    134-148.
Prater, A.J. 1974. The population and migration of Knot in Europe.
    Proc. IWRB Wader Symposium (Warsaw 1973) pp 99-113.

Prater, A.J. 1979. Shorebird census studies in Britain, in
     "Shorebirds in the marine environment" (Studies in Avian Biology
     No. 2) pp 157-166 California: Cooper Ornithological Society.
Ratcliffe, P.J. 1979. An ecological study of the intertidal
     invertebrates of the Humber estuary. Unpublished Ph.D. thesis,
     University of Hull, U.K.
Smith, P.C. 1975. A study of the winter feeding ecology and behaviour
     of the Bar-tailed Godwit Limosa lapponica. Unpublished Ph.D.
     thesis, University of Durham, U.K.
Tasker, M. 1979 in Tasker, M. and Milsom, T.P. Birds of the Humber
     Estuary. Report to Nature Conservancy Council. pp 108.
Townshend, D.J. 1981. The use of intertidal habitats by shorebird
     populations, with special reference to Grey Plover and Curlew.
     Unpublished Ph.D. thesis, University of Durham.

FEEDING AND SURVIVAL STRATEGIES OF ESTUARINE ORGANISMS

EPILOGUE

N.V. Jones
Department of Zoology
University of Hull

To attempt to summarise the forgoing papers seems unnecessary
but it may be appropriate to add some comments of a general nature.

It seems that the subject was rather teasing, particularly to
the authors dealing with invertebrates. The concept of strategies
seems less familiar in this field than it is to the botanists and
ornithologists. Most papers deal primarily with one kind of strategy.
Many deal with the problems of securing sufficient food (McLusky
and Elliott, Wolff et al., the papers on birds) while others discuss
reproduction (Mettam, Hunter) and several are involved with the
physiological stresses expected in estuaries (Joint, Sedgwick,
Wilkinson, Hunter, Pagett). There is also the welcome inclusion
of some physical aspects of estuarine existence (Goulder et al.,
Ratcliffe et al.); a questioning of a familiar behavioural pattern
(Barnes) and an integrative approach to the miniature community of
the meiofauna (Warwick). One paper considers the problems of being
eaten (de Vlas) and one describes the changes brought about by human
activities (Lambeck). Presumably, each species that occurs in an
estuary can be viewed from each of these angles although the relative
importance of each aspect may vary according to the species, its
life style, its distribution, its length of life and its period of
residency in the estuary.

The concept of a strategy implies a choice of alternative
actions. The choice is, of course, not a conscious one but arises
out of the natural selection of the alternatives that best enable
the organism to maintain its population in a particular area. In
view of this, it seems a little surprising that r- and K-selection
have not been mentioned in any paper (although they were raised in
discussion at the meeting). The variable nature of the estuarine

291

environment leads us to expect to find the features associated with
r-selection (low species diversity, small body size, short life-
cycle, early reproduction, rapid development, variable population
size, density independent mortality, low competitive ability and
high production). These features are, no doubt, associated with
the permanent estuarine inhabitants i.e. the organisms that spend
the whole of their lives within the estuary.

The other element of the estuarine fauna is comprised of
periodic exploiters coming into the estuary from outside. These
are the fish and the birds which would be expected to show K-selection
features (large body size, longer life span, slow development, delayed
reproduction, population size fairly constant at or near to the carrying capacity of the environment, mortality more directed, high competitive activity and low production). Will the large, long lived
invertebrates (and algae?) fit into the latter category?

The authors of the papers presented here have elected to confine
themselves to one aspect or another of the life-strategies represented
by the r- and K-selection concepts. This is quite natural as the
title of the meeting invited such a restriction, particularly with
regard to feeding strategies. It is, however, clear that few people
seem to be approaching estuarine organisms from the overall life-
strategy standpoint. We seem to be studying reproduction and production of the resident organisms but concentrating on feeding strategies of the non-resident animals. This is logical as the latter
mainly use the estuary for feeding purposes but it is also possible
that the ease (relative!) of studying such strategies in vertebrates
has something to do with our choice of approach. It may well be that
even the lowly invertebrates have a subtle strategy for maximising
their use of the available food supply, if we only knew how to
recognise it. The large mobile animals have a real choice at the
individual level whereas resident organisms have much less scope
for variability at this level and the choice may be exercised at
the population level. Each aspect obviously has its relevance but
they should be seen as part of the whole life-strategy of the
species.

Some authors have tackled the subject by starting from the
title. I applaud these gallant writers as they have produced useful discussions and reviews. The other authors have started from
their own studies and tried to relate them to the title. Although
this approach is more cautious it has produced a series of interesting and different papers. Together, the two approaches have given
rise to a collection of papers that exphasise that the relative
importance of chemical, physical and biological factors vary
according to the type of organism (phylum), the species, the site,
the time and even the prevailing conditions (short term). This,
of course, will not surprise any ecologist! With such a diversity
of detail to consider, is it not logical and timely to seek an

integrative approach? We usually simplify diversity by classification
and two suitable integrative and functional classifications seem
appropriate: the community (trophic levels, energy flow) and life-
strategies (r- and K-strategies).

This meeting has paid attention to the former (at least in
part) but not to the latter. I suggest that estuarine biology is at
a stage where such an approach may be possible and maybe discussion
at such a level would be appropriate for a future meeting.

INDEX

(Page numbers underlined indicate Figures; with suffix 't' entry occurs in a Table)